MB121

AF405418

MICROBIOLOGY

(PAPER-I)
BACTERIAL CELL AND BIOCHEMISTRY

[2 Credits]

For
F.Y.B.Sc. (Semester - II)
As Per New Revised Syllabus, CBCS Pattern
From June 2019

Dr. Rajashree Bhalchandra Patwardhan
M.Sc., M.Phil, Ph.D., (Microbiology),
Associate Professor, Department of Microbiology,
Haribhai V. Desai College of Arts, Science and Commerce
Pune 411 002.

Dr. Pragati Sunil Abhyankar
M.Sc., Ph.D., SET
Associate Professor, Department of Microbiology,
Haribhai V. Desai College of Arts, Science and Commerce
Pune 411 002.

N5042

Microbiology (Paper - I)

ISBN 978-93-89686-92-0

First Edition : January 2020
© : Authors

Published By:
NIRALI PRAKASHAN
Abhyudaya Pragati, 1312, Shivaji Nagar
Off J.M. Road, PUNE – 411005
Tel - (020) 25512336/37/39, Fax - (020) 25511379
Email : niralipune@pragationline.com

➢ DISTRIBUTION CENTRES

PUNE

Nirali Prakashan : 119, Budhwar Peth, Jogeshwari Mandir Lane, Pune 411002,
(For orders within Pune) Maharashtra, Tel : (020) 2445 2044, Mobile : 9657703145
Email : niralilocal@pragationline.com

Nirali Prakashan : S. No. 28/27, Dhayari, Near Asian College Pune 411041
(For orders outside Pune) Tel : (020) 24690204; Mobile : 9657703143
Email : bookorder@pragationline.com

MUMBAI

Nirali Prakashan : 385, S.V.P. Road, Rasdhara Co-op. Hsg. Society Ltd.,
Girgaum, Mumbai 400004, Maharashtra;
Mobile : 9320129587 Tel : (022) 2385 6339 / 2386 9976,
Fax : (022) 2386 9976
Email : niralimumbai@pragationline.com

➢ DISTRIBUTION BRANCHES

JALGAON

Nirali Prakashan : 34, V. V. Golani Market, Navi Peth, Jalgaon 425001,
Maharashtra, Tel : (0257) 222 0395, Mob : 94234 91860;
Email : niralijalgaon@pragationline.com

KOLHAPUR

Nirali Prakashan : New Mahadvar Road, Kedar Plaza, 1st Floor Opp. IDBI Bank,
Kolhapur 416 012, Maharashtra. Mob : 9850046155;
Email : niralikolhapur@pragationline.com

NAGPUR

Nirali Prakashan : Above Maratha Mandir, Shop No. 3, First Floor,
Rani Jhanshi Square, Sitabuldi, Nagpur 440012, Maharashtra
Tel : (0712) 254 7129;
Email : niralinagpur@pragationline.com

DELHI

Nirali Prakashan : 4593/15, Basement, Agarwal Lane, Ansari Road, Daryaganj
Near Times of India Building, New Delhi 110002
Mob : 08505972553, Email : niralidelhi@pragationline.com

BENGALURU

Nirali Prakashan : Maitri Ground Floor, Jaya Apartments, No. 99, 6th Cross,
6th Main, Malleswaram, Bengaluru 560003, Karnataka;
Mob : 9449043034
Email: niralibangalore@pragationline.com

Other Branches : Hyderabad, Chennai

niralipune@pragationline.com | www.pragationline.com

Also find us on www.facebook.com/niralibooks

Preface ...

We are extremely pleased to present this book **F.Y.B.Sc. Microbiology (Paper-I) - Bacterial Cell and Biochemistry** to all the students who have opted for the subject with great interest. The syllabi for F.Y.B.Sc. Microbiology have been revised and modified so as to widen the scope of the subject to be compatible to present developments and needs of the subject. Our effort is to provide the students with the best guidelines in order to help them to achieve the expected outcomes in these changed circumstances. This book covers the entire new and revised syllabus for the first semester of F.Y.B.Sc. (Microbiology) as prescribed by SPPU. The present book includes the following:

- o A complete coverage of the topics for semester-II
- o Simple scientific language for easy understanding of the subject
- o Precise presentation
- o Topic wise explanation
- o Questions at the end of each topic
- o Use of elaborative and appropriate diagrams

We are confident that, this will be extremely helpful for the students to approach the new revised syllabus and give them confidence to successfully face the revised and reformed examination pattern.

We express a deep sense of gratitude to our publishers Mr. Dineshbhai Furia and Mr. Jigneshbhai Furia of Nirali Prakashan for having shown the confidence in us for accomplishment of the task. We appreciate Dr. G. S. Gugale, our colleague for having boosted us to take up this work and moreover to take it to completion. We are also thankful to our colleague, Dr. Mrs. Suniti Gore who readily gave her expert suggestions while finalizing the book. Our sincere thanks also to all those who directly and indirectly became a part of this.

We have given our best inputs for this book. Any suggestions towards the improvement of this book and sincere comments are most welcome on niralipune@pragationline.com.

With best wishes to all the students who are always our strength.

PUNE   **AUTHORS**

January 2020

Syllabus ...

Credit I: Bacterial Cytology   (15 Hrs.)

- Microbial cell size, shape and arrangements
- Structure, chemical composition and functions of the following components in bacterial cell:
  - Cell wall (Gram positive, Gram negative)
  - Concept of Mycoplasma, Spheroplast, Protoplast, L-form
  - Cell membrane
  - Endospore (spore formation and stages of sporulation)
  - Capsule
  - Flagella
  - Fimbriae and Pili
  - Ribosomes
  - Chromosomal & extra-chromosomal material
  - Cell inclusions (Gas vesicles, carboxysomes, PHB granules, metachromatic granules, glycogen bodies, starch granules, magnetosomes, sulfur granules, chlorosomes)

Credit II: Chemical Basis of Microbiology   (15 Hrs.)

Atom, biomolecules, types of bonds (covalent, co-ordinate bond, non-covalent) and linkages (ester, phospho-diester, peptide, glycosidic)

Chemistry of Biomolecules: Structure, Organization and Functions

Carbohydrates: Definition, Classification

i. **Monosaccharides:** Classification based on aldehyde and ketone groups; structure of Ribose, Deoxyribose, Glucose, Galactose and Fructose.

ii. **Disaccharides:** Glyosidic bond, structure of lactose and sucrose.

iii. **Polysaccharides:** Structure and types

Examples-Starch, Glycogen, Peptidoglycan, Chitin

Lipids: Definition, Classification

i. Simple lipids – Triglycerides, Fats and Oils, Waxes.

ii. Compound lipids – Phospholipid, Glycolipids

iii. Derived lipids – Steroids, Cholesterol

Proteins: Definition, Classification

i) General structure of amino acids, peptide bond.

ii) Types of amino acids based on R group

iii) Structural levels of proteins: primary, secondary, tertiary and quaternary

iv) Study of Hemoglobin, flagellin and cytoskeletal proteins

Nucleic acids: Definition, Classification

i) DNA – structure and composition

ii) RNA – Types (m-RNA, t-RNA, r-RNA), structure and functions.

Classification of Bacteria: Introduction to Bergey's Manual of Determinative and Systemic Bacteriology.

Classification of Viruses: ICTV nomenclature.

Contents ...

Chapter **1**...

Bacterial Cytology

Learning Objectives...

- ➤ To understand about the size, shape and arrangement of bacterial cells.
- ➤ To understand regarding the structures external to cell wall like capsule, flagella, pili, fimbriae.
- ➤ To understand about the cell wall in Gram-positive and Gram-negative bacteria.
- ➤ To understand regarding the structures internal to cell wall like plasma membrane, cytoplasm, ribosomes, inclusion bodies, nuclear material and plasmids

Antony van Leeuwenhoek

Antony van Leeuwenhoek is regarded as the father of microbiology. He is known for the discovery of bacteria. He first observed bacteria in the year 1676 and called them 'animalcules' (from Latin 'animalculum' meaning tiny animal). Using single-lensed microscopes of his own design, he was the first to experiment with microbes. Through his experiments, he was the first to relatively determine their size. Most of the animalcules are now referred to as unicellular organisms.

1.1 INTRODUCTION

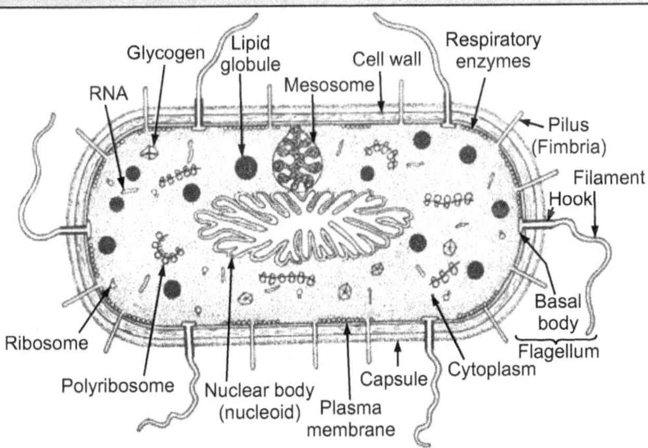

Fig. 1.1

1.1

Bacteria are prokaryotic microorganisms that do not contain chlorophyll. They are unicellular and do not show true branching, except in higher bacteria like actinomycetes. The various structures of bacterial cell differ from one another not only in their physical features but also their chemical characteristics and their functions.

1.2 MICROBIAL CELL SIZE, SHAPE AND ARRANGEMENTS

1.2.1 Size of Bacteria

Bacteria are smaller which can be visualized only under magnification. The unit of measurement used in bacteriology is the micron (Micrometre). Most bacteria are 0.5 to 1.0 μm in diameter. *E. coli*, a bacillus of about average size is 1.1 to 1.5 μm wide by 2.0 to 6.0 μm long. Cyanobacterium *Oscillatoria* is about 7 μm in diameter. The bacterium, *Epulosiscium fishelsoni*, can be seen with the naked eye (600 μm long by 80 μm in diameter). One group of bacteria, called the *Mycoplasmas* measure about 0.25 μ and are the smallest cells known so far. They were formerly known as pleuropneumonia-like organisms (PPLO). *Mycoplasma gallicepticum*, with a size of approximately 200 to 300 nm are thought to be the world smallest bacteria. *Thiomargarita namibiensis* is world's largest bacteria, which is a gram-negative Proteobacterium found in the ocean sediments off the coast of Namibia. Usually it is 0.1-0.3 mm (100-300 μm) across, but bigger cells have been observed up to 0.75 mm (750 μm). Thus, a few bacteria are much larger than the average eukaryotic cell (typical plant and animal cells are around 10 to 50 μm in diameter).

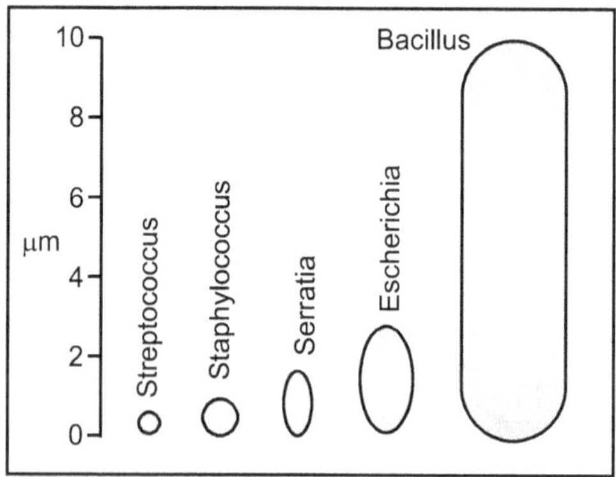

Fig. 1.2: Range of Cell Size in Bacteria

The surface area/volume ratio of bacteria is exceedingly high compared to same ratio for larger organisms of similar shapes. A relatively large surface through which nutrients can enter or waste products leave compared to small volume of cell substance to be nourished accounts for high rate of growth and metabolism in bacteria. Because of high surface area/volume ratio the mass of cell substance to be nourished is very close to the surface, therefore no circulatory mechanism is needed to distribute the nutrients that are taken in.

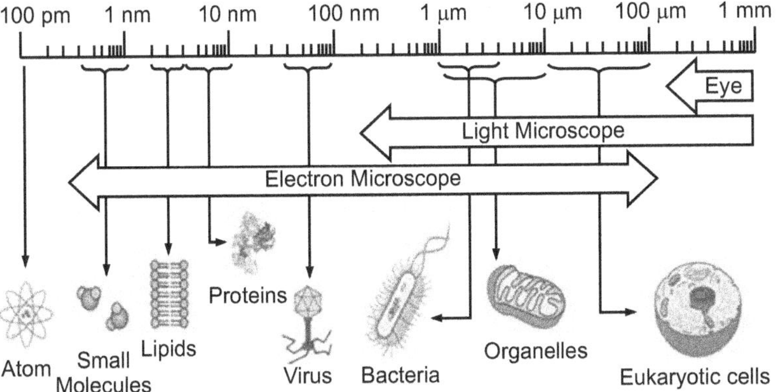

Fig. 1.3

1.2.2 Shapes and Arrangements of Bacteria

The shape of the bacterium is governed by its rigid cell wall. Depending on their shape, bacteria are classified into several types:

1. **Cocci (from kokkos means berry):** These are spherical or oval/ellipsoidal cells. Cocci are round shaped cells which multiply by binary fission and their arrangement depends on the plane of division. The average diameter of spherical bacteria is 0.5-2.0 µm.

   Diplococci: occur in pairs of cells.

   Streptococci: cells arranged in bead or chains.

   Staphylococci: irregular clusters resembling bunches of grapes.

   Sarcinae: cuboidal arrangement of usually 8 or more cells along three dimensions.

   The type of cellular arrangement is determined by the plane through which binary fission takes place and by the tendency of the daughter cells to remain attached even after division.

2. **Bacilli (from baculus meaning rod):** These are rod shaped cells. For rod shaped or filamentous bacteria, length is 1-10 µm and diameter is 0.25-1 .0 µm. They do not have the variety of patterns exhibited by the cocci. Occasionally, they may be:

   Diplobacilli: occurs in pairs.

   Streptobacilli: occur in chains.

3. **Vibrios:** These are comma-shaped, curved rods and derive their name from their characteristic vibratory motility.

4. Some are arranged at angles to each other like cuneiform or **chinese letter pattern (*Corynebacteria*).**

5. **Spirilla:** Spiriila are rigid spiral forms.

6. **Spirochetes** (from spira means coil and chaite means hair): These are flexuous spiral forms. *Spirochaetes* occasionally reach 500 μm in length.

7. **Actinomycetes** (from actis meaning ray and mykes meaning fungus): Actinomycetes are branching filamentous bacteria, so called because of a fancied resemblance to the radiating rays of the sun when seen in tissue.

8. **Mycoplasmas** are the bacteria that are cell wall deficient and hence do not possess stable morphology. They occur as oval bodies and as interlacing filaments.

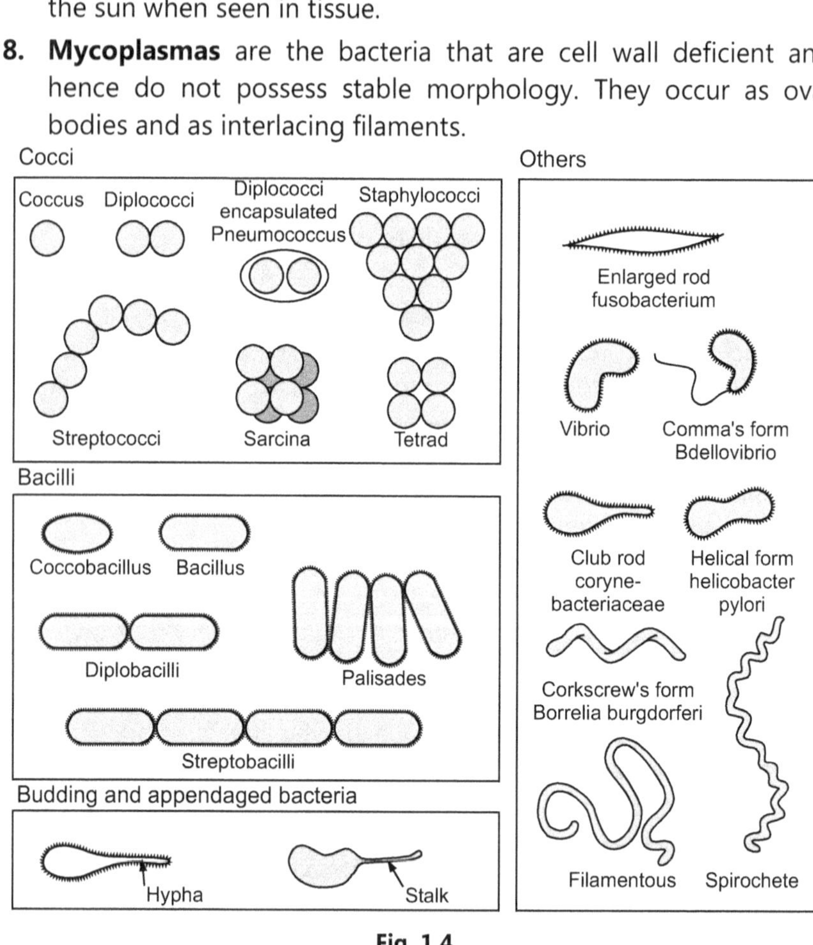

Fig. 1.4

9. When cell wall synthesis becomes defective, either spontaneously or as a result of drugs like penicillin, bacteria lose their distinctive shape. Such cells are called **protoplasts, spheroplasts or L- forms.**

1.3 STRUCTURE, CHEMICAL COMPOSITION AND FUNCTIONS OF THE FOLLOWING COMPONENTS IN BACTERIAL CELL

1.3.1 Cytoplasm

It is the liquid component of bacterial cell consisting of 80% of water and proteins such as enzymes, carbohydrates, lipids, various amino acids, inorganic ions and low molecular weight compounds like organic acids (citric acid, α keto glutarate, succinate), vitamins etc. The major structures present in cytoplasm of bacterial cell are nuclear material, ribosomes and nuclear bodies.

1.3.2 Cell Wall (Gram positive, Gram negative)

The cell wall is the outermost covering of a cell, present adjacent to the cell membrane, which is also called the plasma membrane. The cell wall is present in all plant cells, fungi, bacteria, algae, and some archaea. An animal cell is irregular in their shape and this is mainly due to the lack of cell wall in their cells. There are many variations in chemical composition of the cell wall in prokaryotes, but the basic fundamental structure is consistent in all the microorganisms.

Chemical Composition of Bacterial Cell Wall:

Bacterial cell walls are about 10-25 nm thick and account for about 20-30% of the dry weight of the cell.

Chemically the cell wall is composed of mucopeptide (Peptidoglycan or murein) framework formed by N-acetyl glucosamine and N-acetyl muramic acid molecules alternating in chains, which are cross linked by peptide chains.

The cell wall of Gram-positive bacteria has a simpler chemical nature than those of Gram-negative bacteria.

The cell wall carries bacterial antigens that are important in virulence and immunity.

Gram-positive bacteria
Gram-negative bacteria
Outer membrane
Peptidoglycan
Cell wall
Peptidoglycan
Plasma membrane
(a)
Teichoic acid
Surface protein
Lipoteichoic acid
Peptidoglycan
Cell membrane
Phospholipid
(i) Gram-positive bacteria
Lipopolysaccharide (LPS)
Surface protein
Outer membrane
Lipoprotein
Peptidoglycan
Cell membrane
(ii) Gram-negative bacteria
(b)

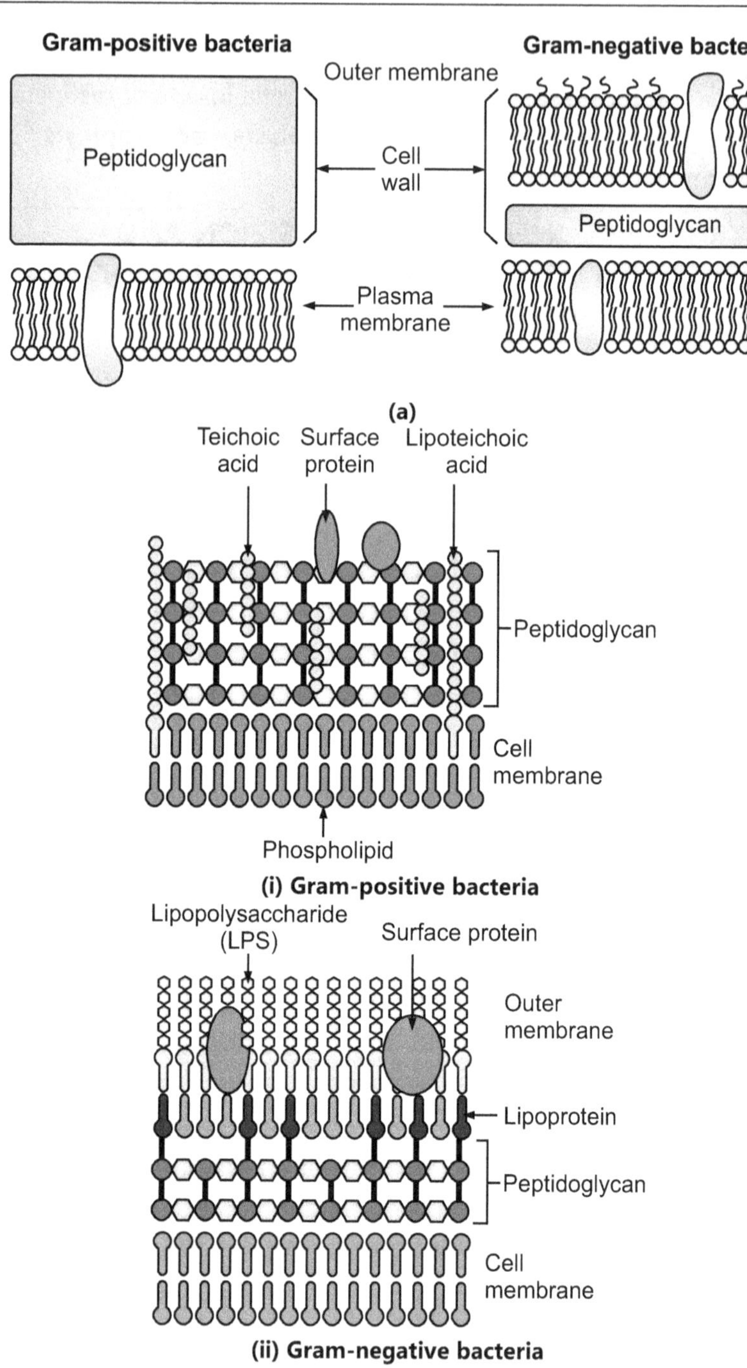

Fig. 1.5

Cell wall of Gram-positive bacteria:

It is homogenous and thick, about 20 – 80 nm in thickness. It mainly consists of peptidoglycan, teichoic acids and sometimes other polysaccharides. Peptidoglycan accounts for about 90% of total dry weight of cell wall.

Cell wall of Gram-positive bacteria consists of the following layers:

1. Peptidoglycan layer: Peptidoglycan is a complex heteropolysaccharide porous cross-linked polymer which is responsible for strength of cell wall. This is thicker (15 - 80 nm) than that of Gram-negative bacteria (2 - 3 nm)

Peptidoglycan is composed of three components.

(a) Glycan backbone: Glycan backbone is the repeated unit of N-acetyl muramic acid (NAM) and N-acetyl glycosamine (NAG) linked by β-1,4 glycosidic bond.

(b) Tetra-peptide side chain (chain of 4 amino acids) linked to NAM: The glycan backbone is cross linked by tetra-peptide linkage. The tetra-peptides are only found in NAM.

Although the peptidoglycan chemistry varies from organism to organism the glycan backbone i.e. NAG-NAM is same in all species of bacteria.

The amino acids found in tetra-peptide are:

1. L-alanine: First position in both Gram positive and Gram-negative bacteria

2. D-glutamic acid: Second position

3. D-aminopimelic acid/ L-lysine: Third position (variation occurs)

4. D-alanine: Fourth position.

(c) Peptide cross linkage:

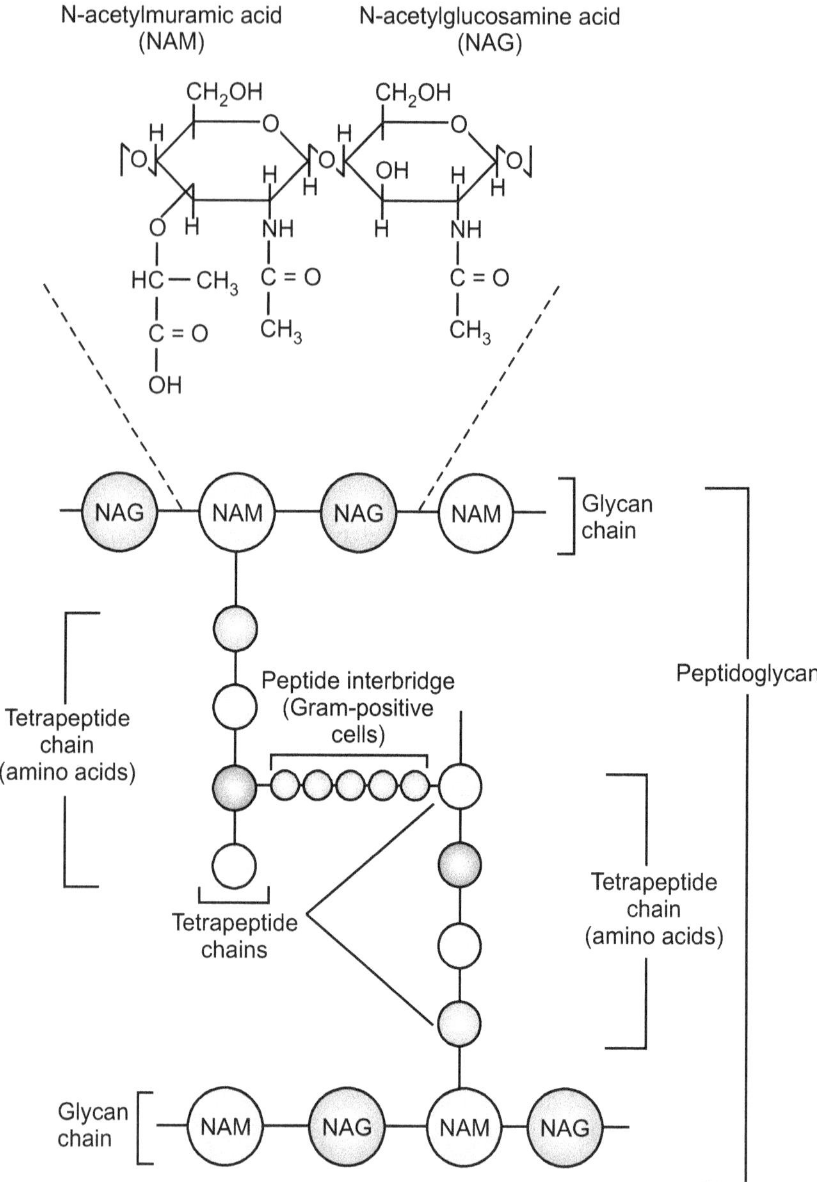

Fig. 1.6

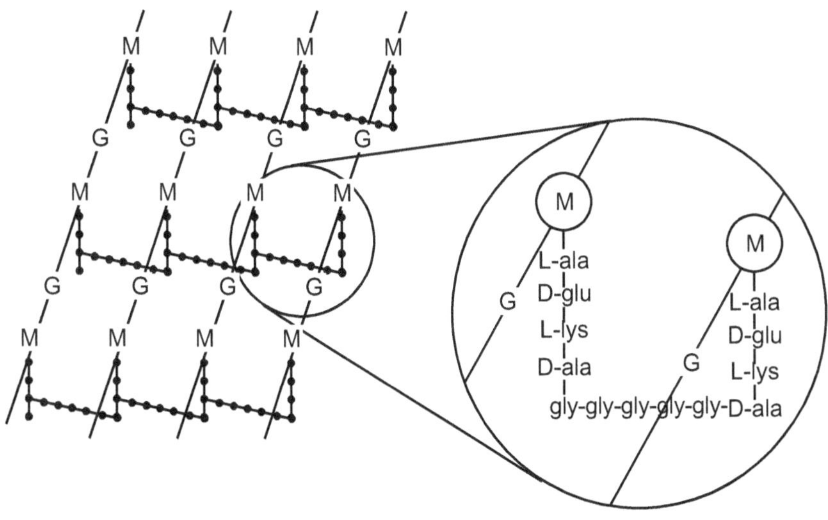

Fig. 1.7

There are many variations in cross linking of different organisms. Based on this the classes of mureins are as follows:

(i) **Group A:** It shows pentaglycine bridge between Meso diaminopimelic acid of one tetrapeptide and D alanine of other.

(ii) **Group B:** It shows pentaglycine bridge between D glutamate of one tetrapeptide and D alanine of other.

(iii) **Group C:** It shows pentaglycine bridge between L lysine of one tetrapeptide and D alanine of other.

2. Teichoic acid layer: Teichoic acid is a major surface antigen of Gram-positive bacteria. Teichoic acid is water soluble polymer of glycerol or ribitol phosphate.

It is present in gram positive bacteria. It binds to protons and maintains the structure of cell wall. It binds with metal ions. It acts as receptors for many viruses. It constitutes about 50% of dry weight of cell wall. It is composed of lipid bilayer, protein and lipopolysaccharide (LPS) layer.

Two types of teichoic acids are present:

- **Cell wall teichoic acid** which is linked to peptidoglycan.
- **Membrane teichoic acid** is also known as lipoteichoic acid which is linked to membrane glycolipid.

Examples of Gram-positive bacteria: *Micrococcus, Staphylococcus, Streptococcus, Leuconostoc.*

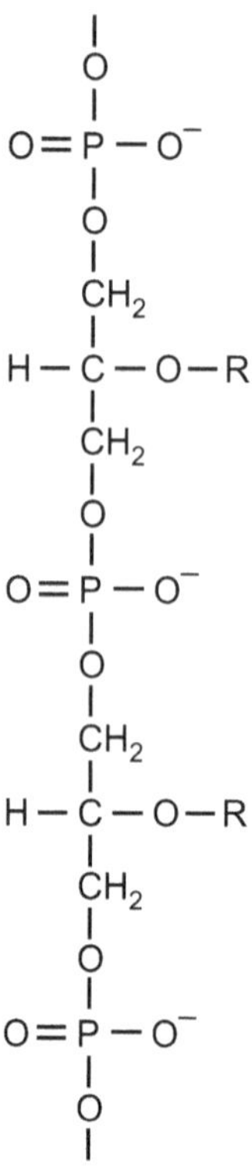

Fig. 1.8: Teichoic Acid Structure

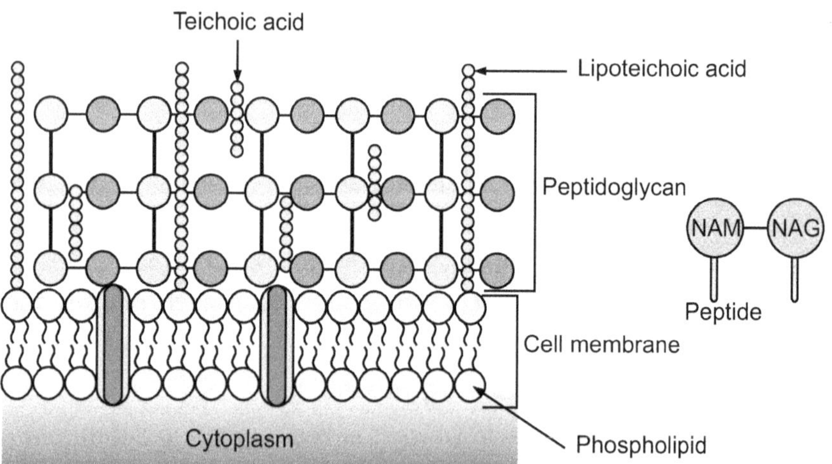

Fig. 1.9: Gram Positive Bacteria Cell Wall

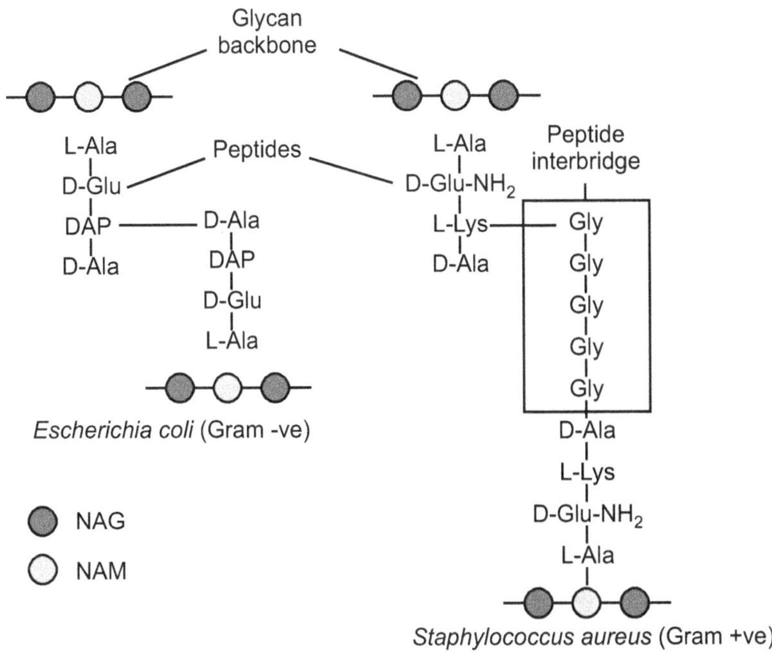

Fig. 1.10

Cell wall of Gram-negative Bacteria:

Cell wall of Gram-negative bacteria is very complex in structure. It is composed of two layers:

1. **Inner layer:** It lies next to cell membrane and is very thin about 2 – 3 nm dimension. It contains peptidoglycan which accounts for 5 – 10% total dry weight of cell and forms a gel like structure. Inner later is separated from outer later by periplasmic space. The periplasm contains high concentration of degradative enzymes and transport proteins. Gram negative cell wall does not contain teichoic acid. Due to less percentage of peptidoglycan Gram negative cells are more susceptible to mechanical damage but less susceptible to lysozyme action.

2. **Outer later:** The outermost layer of Gram-negative bacteria cell wall is called the outer membrane, which contains various proteins known as outer membrane proteins (OMP). Porins are one of them which form transmembrane pores that serve as diffusion channels for small molecules. This is a bilayered membrane containing phospholipid molecules facing polar ends outside and non-polar ends inside the layer. It is in contact with

peptidoglycan present in inner membrane through proteins. This layer contains proteins and lipopolysaccharides.

3. **Lipopolysaccharides** (LPS): LPS are involved in their endotoxic activity and antigenicity. It creates negative charge on the cell. The LPS consist of three regions:

   i. **Glycolipid portion (lipid A):** This region contains glucosamine sugar. It is embedded in outer membrane and contains three fatty acid chains and phosphate or pyrophosphate moieties. It is responsible for the endotoxic activities like pyrogenicity, lethal effect, tissue necrosis, anticomplementary activity, B cell mitogenicity, immunoadjuvant property, antitumor activity etc.

   ii. **Core polysaccharide:** It is joined to lipid A region and contains unusual sugars like 2-Keto-3-deoxy-D-manno-octonic acid (KDO) (a unique eight-carbon sugar), galactose and n-acetyl glucosamine (NAG).

   iii. **O-polysaccharide:** This portion determines O antigen specificity. It is a short polysaccharide chain extending outside from core region. It contains galactose, mannose, rhamnose, fucose, xylose, ribose etc. This 'O' side chain shows variations in the composition in different organisms and thus is very specific.

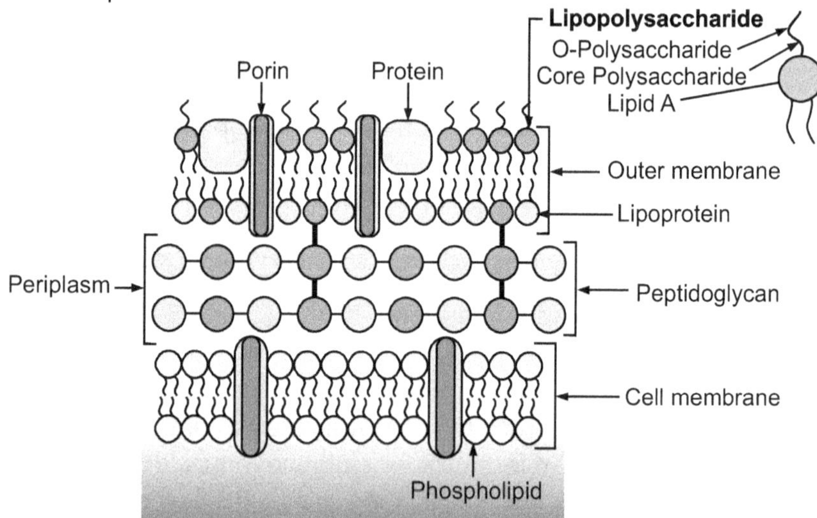

Fig. 1.11

Examples of Gram-negative bacteria: *E. coli, Salmonella, Klebsiella, Shigella, Pseudomonas.*

Functions of the Cell Wall:

1. The cell wall provides definite shape, strength, rigidity to the cell.
2. It also provides protection against mechanical stress and physical shocks.
3. It plays an important role in cell division.
4. It helps to control cell expansion due to the intake of water.
5. It also helps in preventing water loss from the cell.
6. It is responsible for transporting substances between and across the cells.
7. It acts as a barrier between the interior cellular components and the external environment.
8. It protects the cell from toxic substances.
9. It enhances pathogenicity by making cell resistant to engulfment by macrophages.

1.3.3 Concept of Mycoplasma, Spheroplast, protoplast, L-form

a. Mycoplasma:

These are the smallest free-living microorganisms. Mycoplasmas lack cell wall and so are highly pleomorphic with no fixed shape or size. They lack cell wall precursors like muramic acid and diaminopimelic acid.

Morphology of Mycoplasma:

They occur as granules and filaments of various sizes. Granules may be coccoid, balloon, disc, ring, or star forms. The filaments are slender of varying length and slow true branching. Mycoplasmas do not possess spores, flagella or fimbria. Some species exhibit gliding motility. Mycoplasmas are Gram negative but are better stained with Giemsa stain. Multiplication is by binary fission, but as genomic, replication and cell division are often asynchronous, budding forms and chains of beads are produced. A distinct feature seen in some species is a bulbous enlargement, with a differentiated tip structure, by means of which the organisms can adhere to the suitable host cells.

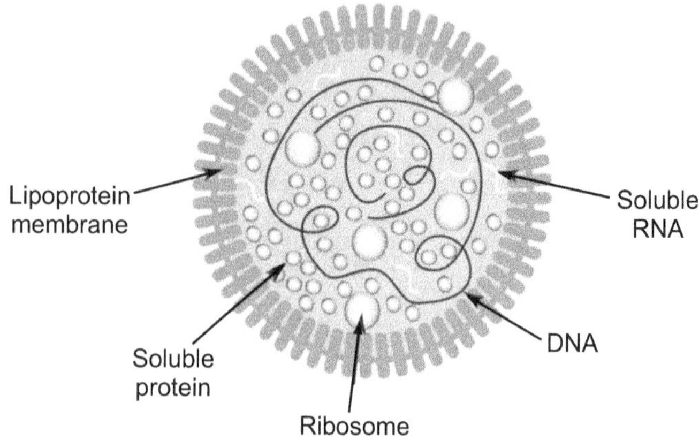

Fig. 1.12

Mycoplasma may be cultivated in fluid or solid media. They are generally facultative anaerobes, growth being better aerobically. They grow within a temperature range of 22-41°C.Colonies appear after 2 to 6 days incubation and are about 10-600 micrometre in size. The colony is typically biphasic with a "fried egg" appearance, consisting of a central opaque granular area of growth extending into the depth of the medium, surrounded by a flat, translucent peripheral zone.

(a) Spheroplast: When lysozyme acts on Gram negative bacteria, usually the wall is not destroyed to the same extent as in gram-positive cells; some of the outer membrane also remains. Lysozyme breaks glycosidic linkage between sugars NAM and NAG, destruct the peptidoglycan layer.   In this case, the cellular contents, plasma membrane and remaining outer wall layer are called a spheroplast. Spheroplasts are spherical in structure. These are the cells lacking only peptidoglycan layer.

(b) Protoplasts: The cellular contents that remain surrounded by the plasma membrane may remain intact if lysis does not occur; this wall-less cell is termed as a protoplast. Protoplast is spherical and is still capable for carrying on metabolism.

(c) L-forms: Kleinberger in 1935 found pleuropneumonia-like forms in a culture of *Streptobacillus moniliformis* and termed as L-forms, after Lister Institute, London, where the observation was made.

It was subsequently shown that many bacteria, either spontaneously or induced by certain substances like penicillin, lost part or all of their cell wall and develop into L forms. Such L – forms may be 'unstable' when they revert back to their normal morphology or stable when they continue in the cell wall deficient state permanently.

Cell wall deficient forms (L-form, protoplasts, spheroplasts) may not initiate disease but may be important in bacterial persistence during antibiotic therapy and subsequent recurrence of the infection.

1.3.4 Cell Membrane

It is also known as cytoplasmic membrane. It is a thin (5-10 nm) layer lining the inner surface of cell wall and separating it from cytoplasm. It is a limiting membrane in which the cell contents are enclosed. It acts as a semipermeable membrane regulating the flow and exchange of metabolites between the cell interior and its environment. It receives signals, coordinates molecular interactions, adhesion and communication.

Structure of Cell Membrane:

Plasma membrane is two layered structure. Chemically, the membrane consists of phospholipids and proteins. Phospholipids account for 20-30% of dry weight of membrane and proteins account for 60-70% of dry weight of membrane.

To explain the way in which the phospholipids and proteins are arranged various models have been proposed. The widely accepted model is '**Fluid Mosaic Model**' which is given by S.J. Singer and Garth L. Nicolsonin 1972. This model postulated that the Phospholipids form a bilayer in which proteins are embedded. The proteins form a mosaic like structure in the phospholipid bilayer. The lipids are amphoteric. Each phospholipid molecule contains a polar head, composed of phosphate group and glycerol that is hydrophilic (water-loving) and soluble in water, and non-polar tails, composed of fatty acids that are hydrophobic (water-fearing) and insoluble in water. The polar heads are on the outer surfaces of the lipid bilayer, and the non-polar tails are in the interior of the bilayer. The individual phospholipid molecules can move laterally giving the bilayer fluidity, flexibility, high electric resistance and relative impermeability to highly polar molecules.

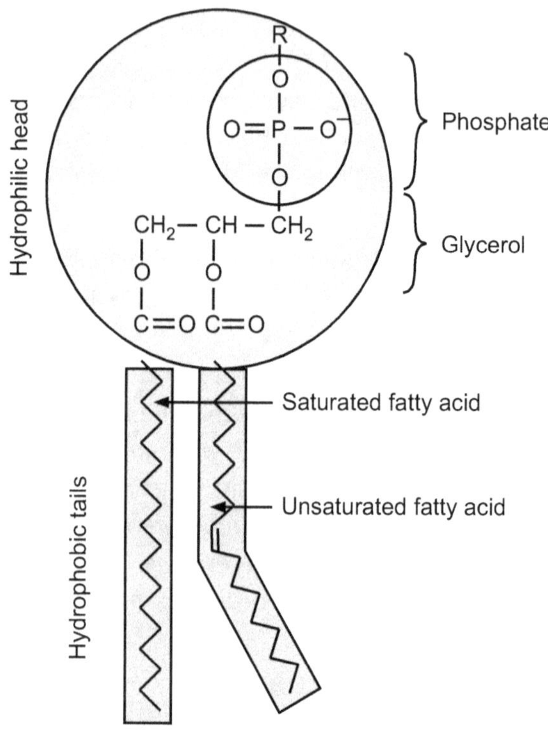

Fig. 1.13

The membrane protein molecules are globular and arranged in a variety of ways. Some, called **peripheral proteins**, are easily removed from the membrane by mild treatment and lie at the inner or outer surface of the membrane. They function as enzymes that catalyse chemical reactions, as a "scaffold" for support, and as a mediator of changes in membrane shape during movement.

Other proteins, called **integral proteins**, can be removed from the membrane only after disrupting the lipid bilayer. Most integral proteins penetrate the membrane completely and are called **transmembrane proteins**. Some integral proteins are channels that have a pore through which substances enter and exit the cell.

Many of the proteins and some of the lipids on the outer surface of the plasma membrane have carbohydrates attached to them. Proteins attached to carbohydrates are called **glycoproteins**; lipids attached to carbohydrates are called glycolipids. Both glycoproteins and glycolipids help to protect and lubricate the cell and are involved in cell-to-cell interactions.

Due to rapid movement of lipid and protein molecules the membrane is quasi-fluid structure. It acts like a solid by enclosing the cell contents (something that a fully fluid liquid could not do) and like a liquid by allowing molecules to move through it by osmosis and diffusion.

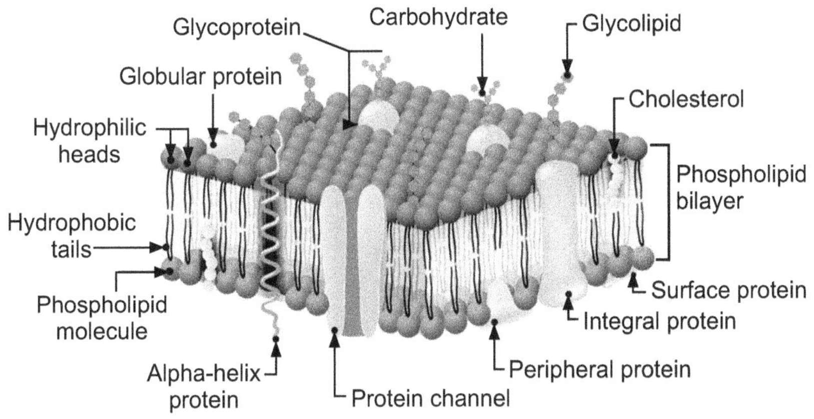

Fig. 1.14: Cell membrane

Functions of Cell Membrane:

1.   The most important function of the plasma membrane is to serve as osmotic barrier of the cell involved in endoosmosis and exoosmosis depending on surrounding hypotonic or hypertonic solution. Plasma membranes have selective permeability which allows only certain molecules and ions to pass through the membrane.

Transport mechanisms across the membrane:

Cell membrane is selectively permeable. It interacts with the environment by selective transport of nutrients inside the cell. Transport mechanisms are as follows.

i.   **Simple diffusion:** Movement of small non-polar molecules across the membrane from region of high concentration of solute to low concentration of solute down the gradient. It is also known as passive diffusion. E.g. Transfer of O_2 and CO_2 gases.

ii. **Osmosis:** It is diffusion of water across semipermeable membrane. It allows movement of water from higher concentration to lower concentration of water. Osmosis can result in two processes.

- **a. Plasmolysis:** Shrinkage of cell due to loss of water in the cell when placed in hypertonic solution.

- **b. Plasmoptysis:** In hypotonic solution water enters the cell. Cell swells and bursts or lyse.

iii. **Facilitated diffusion:** It is the transport of molecules down the gradient at higher rate as some carrier proteins are embedded in cytoplasmic membrane. When solute concentration is high outside, this mechanism causes invert movement of solute but revert mechanism does not occur.

- a. **Uniport:** This allows movement of single solute particle.

- b. **Co-transport:** The movement of two molecules takes place.

- (1) **Symport:** Both solute molecules are transported in the same direction.

- (2) **Antiport:** Direction of movement of two solutes molecules is opposite. One is taken inside the cell and other is transported outside the cell.

iv. **Active transport:** (Uphill transport) It is the movement of solute molecules against the concentration gradient by using either energy obtained from ATP hydrolysis or by transport of ions. Active transport is like pump mediated by carrier proteins. Depending on energy source used for the transport there are different types.

- **a. Direct active transport:** ATP driven transport. The energy released during hydrolysis of ATP is used for transporting solute. It is common in Gram negative organisms like *E. coli.*

- **b. Indirect active transport:** It depends on the co-transport of ions. These ions may be Na^+, H^+. e.g. Na^+ K pump.

v. **Group translocation:** Solute is chemically modified in order to facilitate movement of PEP sugar, phosphotransferase system in which sugar is phosphorylated and then enters inside the cell.

2. Plasma membrane is a semipermeable and concerned with active absorption of minerals. It concentrates nutrients in the cell and excretes waste products.

3.   It is involved in the formation of cell organelles and hence also called as unit membrane.

4.   It is concerned with ingestion of food particles i.e. endocytosis, a process by which foreign food particles are taken in and are digested. Endocytosis may be (i) Pinocytosis: Ingestion of liquid food, (ii) Phagocytosis: Ingestion of solid food.

5.   In aerobic bacteria electron transport chain is in plasma membrane so it is mitochondrial equivalent.

6.   In purple bacteria photosynthetic apparatus is located in the plasma membrane.

7.   Plasma membranes are also important to the breakdown of nutrients and the production of energy.

8.   The plasma membranes of bacteria contain enzymes capable of catalysing the chemical reactions that break down nutrients and produce ATP.

9.   In some bacteria, pigments and enzymes involved in photosynthesis are found in infolding of the plasma membrane that extend into the cytoplasm.

10. Plasma membrane pores are 8 - 10 A° wide lined with hydrophilic residues which allows addition of water molecules to surrounding medium.

11. Plasma membrane anchors the DNA during replication.

1.3.5 Endospore (Spore Formation and Stages of Sporulation)

Spores are unicellular structures. They are highly resistant, dormant metabolically inactive resting structures formed in response to adverse environmental conditions such as lack of nutrients, environmental stress etc. Spore is a refractile body formed by a vegetative mother cell. They do not have a role in reproduction. Vegetative cells of endospore forming bacteria begin sporulation when a key nutrient, such as the carbon or nitrogen source, becomes scarce or unavailable. Endospores are highly durable dehydrated cells with thick walls and additional layers. Endospores enable bacteria to lie dormant for extended periods, even centuries. There are many reports of spores remaining viable over 10,000 years.

There are two types of spores:

a. Endospores – Form inside the vegetative cell.

b. Exospores – Form in either one of the ends of the vegetative cell.

The diameter of the endospore may be the same as, smaller than, or larger than the diameter of vegetative cell. Depending on species, the endospore might be located terminally (at one end), sub-terminally (near one end), or centrally inside the vegetative cell. Bacteria show central bulging or non-bulging spores. These are highly resistant to chemical disinfectants, low or high temperature as well as radiations. Sporulating bacteria includes *Bacillus subtilis*, *Bacillus polymyxa*, *Bacillus licheniformis*, *Clostridium*, *Sporolactobacillus*, *Thermoactinomycetes*,

Structure of Endospore:

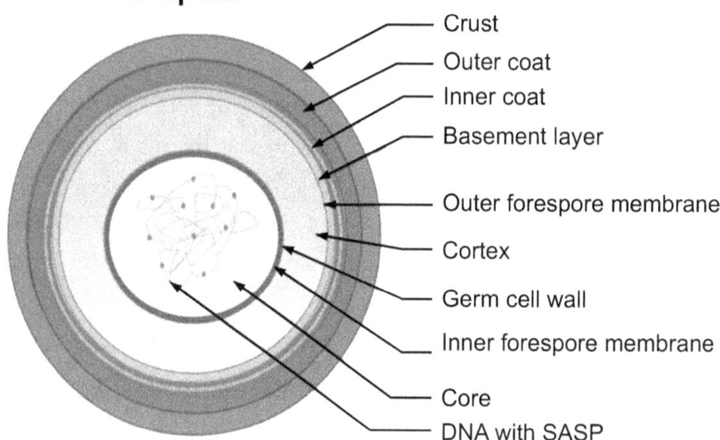

SASP: small, acid-soluble spore proteins

Fig. 1.15

Components of Endospores:

1. **Core/Protoplast:** The core is the spore protoplast. It contains a complete nucleus (chromosome), all the components of the protein-synthesizing apparatus including ribosomes, and an energy-generating system based on glycolysis. Several unique enzymes are formed (e.g., dipicolinic acid synthetase). Spores contain no ATP. The energy for germination is stored as 3-phosphoglycerate rather than as ATP. It takes up large amount of Ca ions which chelates with dipicolinic acid to form Calcium dipicolinate.

2. **Spore germ cell wall:** The innermost layer surrounding the inner spore membrane is called the spore wall. It contains normal peptidoglycan and becomes the cell wall of the germinating vegetative cell.

3. **Cortex:** The cortex is the thickest layer of the spore envelope, accounts for half volume of spore. It has inner and outer cortex membrane. It contains modified form of peptidoglycan. Cortex peptidoglycan makes the spore resistant.

4. **Coat:** Around the cortex a layer of spore coat is present which contains proteins with high percentage of cysteine, keratin and hydrophobic amino acids. Spore coat contains two layers inner and outer coat. The impermeability of this layer gives spores their relative resistance to antibacterial chemical agents.

5. **Exosporium:** In some bacteria outer loose envelope surrounds endospore known as exosporium. The exosporium is a lipoprotein membrane containing some carbohydrates.

Sporulation Process:

The process of endospore formation within a vegetative cell takes several hours and is known as sporulation or sporogenesis.

Stage I: Axial filament formation stage

- Sporulation process starts with unusual event of formation of axial filament of nuclear material which is usually present in dispersed form. Bacterial chromosome becomes thread like known as axial filament.
- Axial filaments attached to cytoplasmic membrane by mesosome.
- Elongation of cell take places.
- PHBA is the reserved food material in *Bacillus spp.* is utilized in sporulation.
- In the first stage of sporulation, a newly replicated bacterial chromosome and a small portion of cytoplasm are isolated by an ingrowth of the plasma membrane called a spore septum.

Stage II: Forespore formation

- After formation of axial filament unequal division of cell takes place due to inward folding of cytoplasmic membrane at one pole which is known as **invagination**.

- The spore septum becomes a double layered membrane that surrounds the chromosome and cytoplasm. This structure, entirely enclosed within the original cell, is called a forespore.
- Most of the water present in the forespore cytoplasm is eliminated during completion of sporulation.

Stage III: Engulfment of forespore

- Mother cell membrane grows around the forespore engulfing it.
- Forespore derives one more membrane called outer core wall.
- Thick layers of peptidoglycan are laid down between the two membrane layers.

Stage IV: Synthesis of exosporium

- Chromosome of mother cell disintegrates.
- Exosporium synthesis occurs.
- Forespore starts forming primodial cortex between two membranes which fills gap between two membranes.
- Dehydration of cell.

Stage V: Synthesis of dipicolonic cacid

- Production of SASPs (small acid-soluble spore proteins) and dipicolinic acid occurs.
- Incorporation of calcium ions with dipicolonic acid occur forming calcium dipicolonate.
- Further dehydration of cytoplasm. At this stage metabolic activity is very less as compared to vegetative cell and spore starts appearing as refractile body.
- Multi-layered lamellar structure fuses to form thick spore coat. Deposition of spore coat proteins and accumulation of cystine leads to completion of spore coat structure. This coat is responsible for the resistance of endospores to many harsh chemicals.

Stage VI: Maturation

- At this stage cortical peptidoglycan synthesis continues forming more homogenous protoplast. Protoplast becomes electron dense structure and spore coat synthesis gets completed which makes it heat resistant and refractile.

Stage VII: Release of endospore

- When the endospore matures, the vegetative cell wall ruptures, and original cell is degraded, and release of endospore takes place.
- The released highly dehydrated endospore core contains only DNA, small amount of RNA, ribosomes, enzymes and a few important small molecules and contain large amount of organic acid called dipicolinic acid (found in cytoplasm), which is accompanied by large number of calcium ions. These cellular components are essential for resuming metabolism later.

Spore Germination:

Bacteria remains dormant by forming spores for years. This dormancy may be retained for years but this condition may be broken up due to change in the environmental conditions. Under favourable conditions each endospore germinates to give rise to a vegetative cell. Transformation of dormant spore into vegetative cell is called as germination.

The germination process occurs in three stages: activation, initiation, and outgrowth.

- **Activation:** Endospores cannot germinate immediately after they have formed, but they can germinate after they have rested for several days. They need certain conditions to be activated.
- **Initiation:** Once activated, a spore will initiate germination if the environmental conditions are favorable. Autolysin will be activated, and it will rapidly degrade the cortex peptidoglycan. Water is taken up, calcium dipicolinate is released, and a variety of spore constituents are degraded by hydrolytic enzymes.
- **Outgrowth:** Degradation of the cortex and outer layers results in the emergence of a new vegetative cell consisting of the spore protoplast with its surrounding wall. Now, using nutrients around, the cell can multiply again.

Significance of Endospores:

Endospores are important from clinical viewpoint and in the food industry because they are resistant to process that normally kill vegetative cells. Such processes include heating, freezing, desiccation, use of chemicals and radiation. Most vegetative cells are killed by temperatures above 70°C, endospores can survive in boiling water for several hours or more. Endospores of thermophilic bacteria can survive in boiling water for 19 hours.

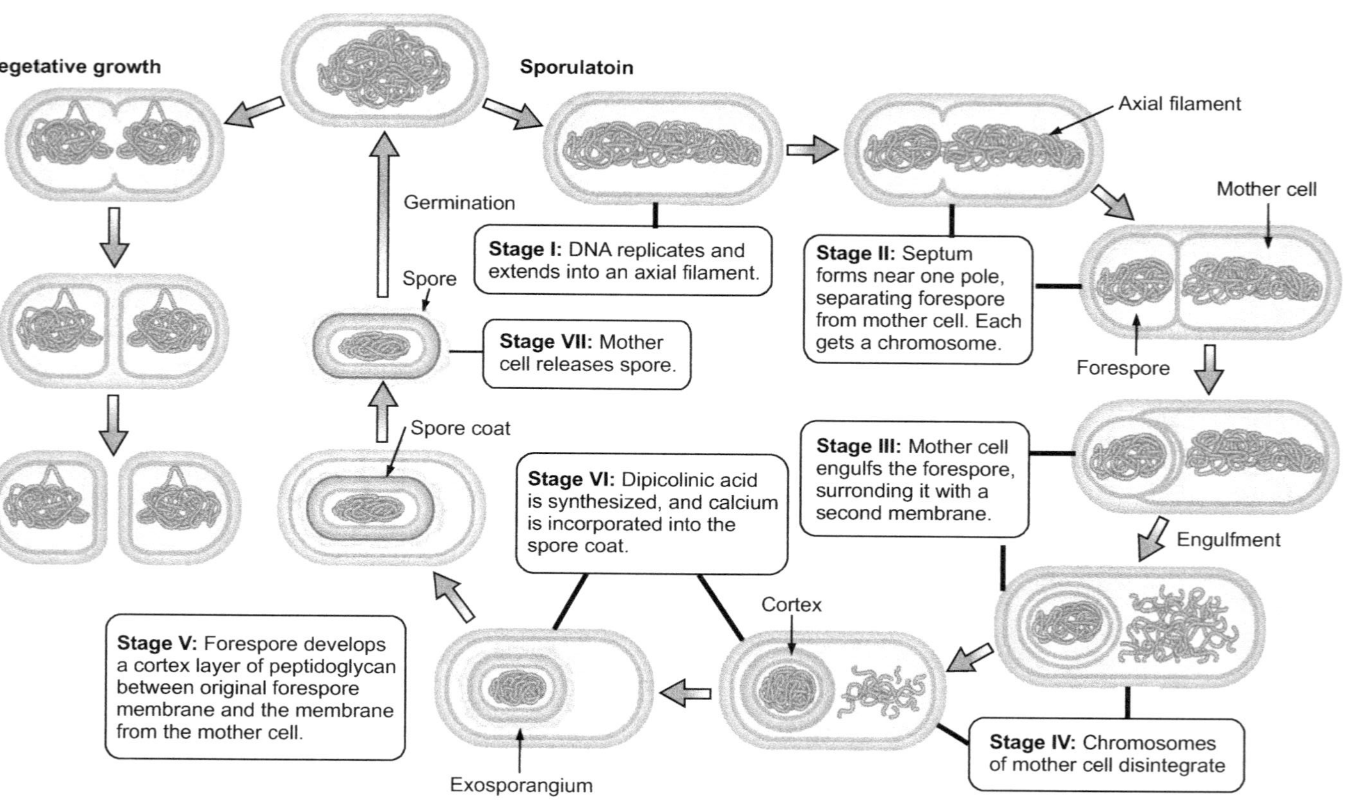

Vegetative growth
Sporulatoin
Axial filament
Mother cell
Germination
Spore
Forespore
Stage I: DNA replicates and extends into an axial filament.
Stage VII: Mother cell releases spore.
Stage II: Septum forms near one pole, separating forespore from mother cell. Each gets a chromosome.
Spore coat
Stage VI: Dipicolinic acid is synthesized, and calcium is incorporated into the spore coat.
Stage III: Mother cell engulfs the forespore, surronding it with a second membrane.
Engulfment
Cortex
Stage V: Forespore develops a cortex layer of peptidoglycan between original forespore membrane and the membrane from the mother cell.
Exosporangium
Stage IV: Chromosomes of mother cell disintegrate

Fig. 1.16

1.3.6 Bacterial Capsule

- The bacterial capsule is a very large structure of many bacteria.
- It is a polysaccharide layer that lies outside the cell envelope and is thus considered as a part of the outer envelope of a bacterial cell.
- It is a well-organized layer, and is the cause of various diseases.
- It is 0.2 μm thick viscous layer firmly attached to the cell wall of some capsulated bacteria.
- It is amorphous gelatinous covering surrounding cells.
- Usually capsulated material is not highly water soluble and don't diffuse outside the cell but sometimes it is highly water soluble and dissolves in the medium increasing viscosity of the broth.
- The capsule is found in both gram negative and gram-positive bacteria. It is different from the second lipid membrane i.e. bacterial outer membrane, which contains lipopolysaccharides and lipoproteins and is found only in gram-negative bacteria.
- Capsule is very delicate structure. It can be removed by vigorous washing.
- Capsule is most important virulence factor of bacteria.
- Capsule in visualized by Negative staining technique.
- If it is thin and cannot be seen by light microscopy then it is known as microcapsule
- If capsule is too thick it is known as slime.
- Slime layer are loosely attached to cell wall and can be lost on vigorous washing and on subculture.

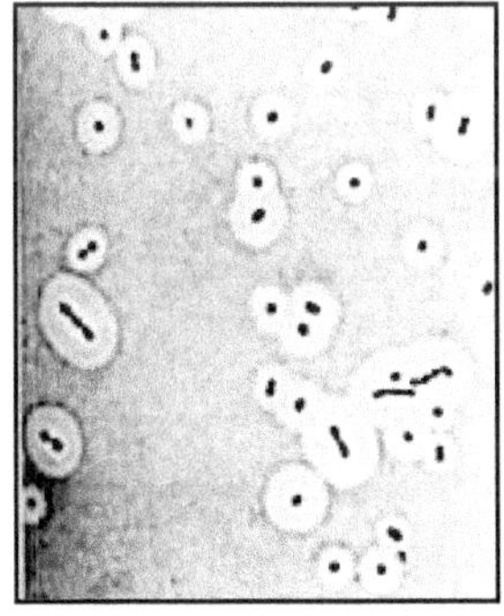

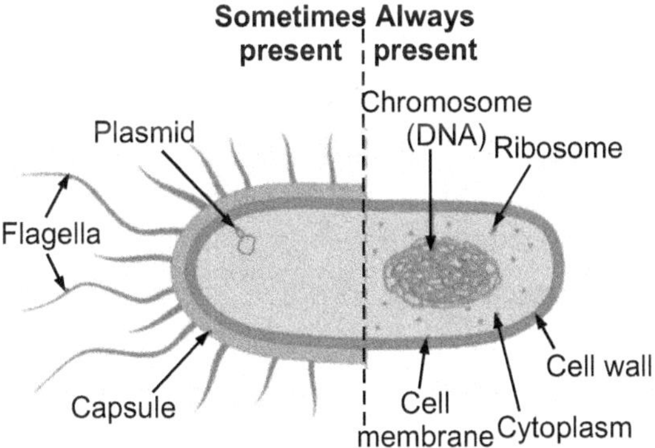

Fig. 1.17

- Capsule and slime layer are sometimes summarized under the term glycocalyx.
- As a group where the capsule is present, they are known as polysaccharide encapsulated bacteria or encapsulated bacteria.

Composition of Capsule:

- 98% water and 2% polysaccharide or glycoprotein/ polypeptide or both.
- It contains usually homopolysaccharides which are synthesized outside the cell from disaccharides by extracellular enzymes. In case of Streptococcus mutans capsule contains glucan which is polymer of glucose made by using sucrose as a sugar source. In case of Acetic acid bacteria, capsule is composed of homopolysaccharide (hemicellulose).
- *Leuconostoc:* Capsule is composed of cellulose, consisting of glucose or fructose.
- Other capsules are composed of heteropolysaccharides containing various sugars.
- *Klebsiella pneumoniae:* Capsule is made up of glucose, galactose, rhamnose etc.
- In *Bacillus anthracis,* capsule is made up of Polypeptide (Polymer of D-glutamic acid) and in Streptococci, it is L-aminoacids.
- *Bacillus megaterium* synthesizes a capsule composed of polypeptide and polysaccharides.
- *Streptococcus pyogenes* synthesizes a hyaluronic acid capsule.
- *Streptococcus agalactiae* produces a polysaccharide capsule of nine antigenic types that all contain sialic acid.

Functions of Capsule:

- **Prevent the cell from desiccation and drying:** Capsular polysaccharide binds with significant amount of water making cell resistant to dry.
- **Protection:** It protects from mechanical injury, temperature, drying etc.
- **Attachment:** Capsule helps to adhere bacteria to smooth surfaces. E.g. *Streptococcus mutans* which cause dental carries attach on teeth surface by its capsule. Polysaccharide capsules

potentially favour the attachment to surfaces and the formation of biofilms through non-specific forces such as electrostatic and van der Waals forces and/or specific ligand-receptor interactions.

- **Anti-phagocytic:** Capsule resist phagocytosis by macrophages.
- Capsule prevent attachment of bacteriophage on cell surface making the cells resistant to bacteriophages
- **Source of nutrition:** Capsule is source of nutrition when nutrient supply is low in cell.
- **Repulsion:** If the capsule is made up of negatively charged compounds like sugar uronic acid they will repel other cells due to same negative charge on bacterial cell wall and thus avoid clumping or aggregation of cells.
- **Increase virulence or pathogenicity:** The capsule is considered a virulence factor because it enhances the ability of bacteria to cause the disease (e.g. prevents phagocytosis).
- The capsule can protect cells from engulfment by eukaryotic cells, such as macrophages.

Examples of Capsulated bacteria:
- *Bacillus subtilis*
- *Bacillus anthracis*
- *Klebsiella pneumoniae*
- *Haemophilus influenzae*
- *Clostridium perfringens*
- *Neisseria meningitidis*
- *Pseudomonas aeruginosa*
- *Acinetobacter calcoaceticus*
- *Bacillus megaterium*
- *Streptococcus pyogenes*
- *Streptococcus pneumoniae* (has at least 91 different capsular serotypes. These serotypes are the basis for the pneumococcal vaccines).
- *Streptococcus agalactiae*
- *Staphylococcus epidermidis*
- *Staphylococcus aureus*
- *Escherichia coli*
- *Salmonella*

1.3.7 Bacterial Flagella

- It is the locomotory organ in bacteria.

- Bacterial flagella are long, thin (about 20 nm), and whip like appendages that move the bacteria towards nutrients and other attractants.

- Flagella are free at one end and attached to the cell at the other end.

- Flagellum can never be seen directly with the light microscope but only after staining with special flagella stains that increase their diameter.

- Motile bacteria except spirochetes, possess one or more unbranched, long, sinuous filaments called flagella which are organ of locomotion.

- The filament is external to the cell and connected to the hook at the cell surface. The hook-basal body portion is embedded in the cell envelope. The hook and basal body are antigenically different.

- The flagella are 3-20 μm long and are of uniform diameter (0.01-0.013 μm) and terminate in the square tip.

- The wavelength and thickness of the filament are characteristics of each species. Some bacteria exhibit biplicity that is they have flagella of two different wavelengths.

- Flagella are made of a protein known as flagellin.

- The presence or absence of flagella and their number and arrangement are characteristic of different genera of bacteria.

Structure of Flagella:

The structure of bacterial flagella has been described by Simon (1978), Doestsch and Sjoblad (1980) and Ferris and Beveridge (1985).

1. **Basal body:** It is composed of central rod inserted into series of rings which is attached to cytoplasmic membrane and cell wall.

In Gram-negative bacteria two pairs of rings, the proximal ring and the distal ring, are connected by a central rod. There are two pairs of rings. The outer pair of rings, L-ring and P-ring are attached to respective polysaccharide and peptidoglycan layer of cell wall, and the inner pair of rings i.e. S-ring and M-ring are attached to cell membrane. Four rings are L-lipopolysaccharide ring,

P-peptidoglycan ring, S-super membrane ring, and M-membrane ring.

L-ring: it is the outer ring present only in Gram negative bacteria, it is anchored in lipopolysaccharide layer.

P-ring: it is second ring anchored in peptidoglycan layer of cell wall.

M-S rings: anchored in cytoplasmic membrane.

The outer rings form a bearing for the rod to pass through it. **In Gram-positive bacteria only the distal (inner) pair of rings is present**. The S-ring is attached to inside thick layer of peptidoglycan and M-ring is attached to cell membrane.

2. Hook: It is the wider region at the base of filament. It connects filament to the motor protein in the base. Length of hook is longer in gram positive bacteria than gram negative bacteria.

3. Filament: It is thin hair like structure arises from hook.

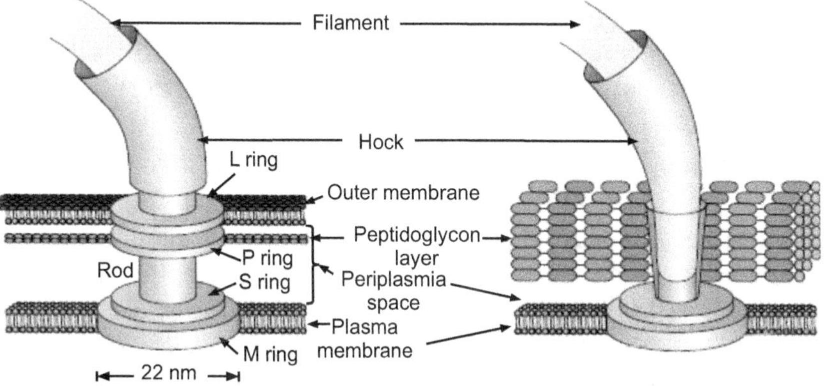

(a) In gram negative bacteria   **(b) In gram positive bacteria**

Fig. 1.18: Structure of Flagella

There are various types of flagellar arrangements:

1. **Monotrichous:** Presence of single flagella in one end of cell. Examples: *Vibrio cholera, Pseudomonas aeruginosa.*

2. **Lophotrichous:** Presence of bundle of flagella in one end of cell. Example: *Pseudomanas fluroscence.*

3. **Amphitrichous:** Presence of single or cluster of flagella at both end of cell. Example; *Aquaspirillium.*

4. **Peritrichous:** Presence of flagella all over the cell surface. Example: *E. coli, Salmonella, Klebsiella.*

5. **Atrichous:** Absence of flagella. Example: Shigella.

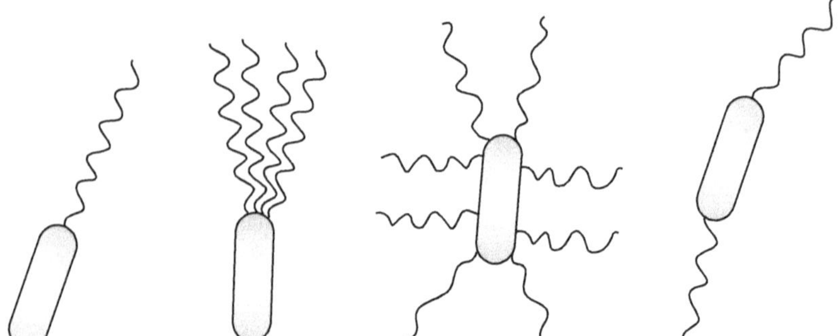

(a) Monotrichous (b) Lophotrichous   (c) Peritrichous   (d) Amphitrichous
   (polar)

Fig. 1.19

Mechanism of Flagellar Motility:

* The movement of flagella results from rotation of basal body which is like the movement of the shaft of an electric motor. Movement of flagella is due to rotational force or 'Torque' created by proton motive force. This proton motive force is generated due to translocation of protons from 'M' ring. This provides chemical gradient i.e. concentration gradient which shows higher concentration of protons outside and less concentration inside the cytoplasm. It also creates pH gradient across the membrane. This chemical gradient provides chemical energy which results in revolution of flagella. Interaction of S and M ring allows movement of hook and filament.

* At the base surrounding the inner ring (M-S and C ring) there is a series of protein called Mot protein. A final set of protein called Fli protein function as motor switch. The flagella motor rotates the filament as a turbine causing movement of the cell in the medium. A turning motion is generated between S-ring and M-ring. S-ring acts as starter while M-ring acts as rotor. The basal body as a whole gives a universal joint to the cell and allows complete rotation of hook and filament. Rotation of flagella is either clockwise or anticlockwise.

Tactic Behaviour:

Chemotaxis: Movement of cell towards or away from chemicals (attractant or repellent respectively).

Phototaxis: Movement towards light.

Aerotaxis: Movement towards oxygen.

Magnetotaxis: Movement towards magnetic field.

Examples of motile and non-motile bacteria

Motile bacteria: *E. coli, Pseudomonas, Salmonella, Bacillus.*

Non-motile bacteria: *Streptococcus, Staphylococcus, Klebsiella pneumoniae.*

1.3.8 Fimbriae and Pili

Many Gram-negative bacteria have hair like appendages which are shorter, thinner, straighter than flagella. They are composed of '**pilin**' proteins arranged helically around the central core. These appendages are divided into two types **Fimbriae and Pili**.

1.3.8.1 Fimbriae

Fimbriae *(sing. fimbria)* are very fine, thin, hair like filamentous appendages that extend from the cell, often in the tens or hundreds and are used by the cell to attach the surfaces.

They can occur at poles of bacterial cell or they can be evenly distributed on the entire surface of the cell.

They can be particularly important for pathogenic bacteria, which use them to attach to host tissues.

They are shorter and thinner than flagella (about 0.5 μm long and less than 10 nm thick) and project from the cell surface as straight filaments.

Fimbriae enable a cell to adhere to surfaces, including surface of other cells. Fimbriae function as organs of adhesion, helping the cells to adhere firmly to particles of various kinds.

Example: Fimbriae of the bacterium *Neisseria gonorrhoeae*, the causative agent of gonorrhoea, help the bacteria to colonize on mucous membranes. This enhances the pathogenicity of bacterium.

1.3.8.2 Pili (sing. pilin)

Many gram-negative bacteria contain hair-like appendages called pili that are shorter, straighter and thinner than flagella and are used for attachment and transfer of DNA rather than for motility. Pili are absent in many Gram-positive bacteria.

Pili are unrelated to motility and are found on motile as well as non-motile cells.

Pili are very similar to fimbriae but are typically longer and fewer in number than fimbriae with only 1-2 present on each cell.

They are thin filamentous appendages that extend from the cell and are made of 'pilin' protein arranged helically around a central core.

Pili can be used for attachment as well, to both surfaces and host cells, such as the *Neisseria gonorrheae* cells that use their pili to grab onto sperm cells, for passage to the next human host.

The **conjugative pili** participate in the process known as **conjugation**, which allows for the transfer of a small piece of DNA from a donor cell to a recipient cell.

The **type IV pili** play a role in an unusual type of motility known as **twitching motility**, where a pilus attaches to a solid surface and then contracts, pulling the bacterium forward in a jerky motion. This enables bacteria to "crawl" or "walk" over the surfaces to which they have attached by extending and retracting their type IV pili. Bacteria with type IV pili include *Pseudomonas aeruginosa, Neisseria gonorrhoeae, Neisseria meningitidis*, and *Vibrio cholerae*.

At the end of the pilus shaft is the adhesive tip structure having a shape corresponding to that of specific glycoprotein or glycolipid receptors on a host cell.

Because both the bacteria and the host cells have a negative charge, pili may enable the bacteria to bind to host cells without initially having to get close enough to be pushed away by electrostatic repulsion. Once attached to the host cell, the pili can depolymerize and enable adhesions in the bacterial cell wall to make more intimate contact.

Bacteria are constantly losing and reforming pili as they grow in the body and the same bacterium may switch the adhesive tips of the pili in order to adhere to different types of cells and evade immune defenses.

There are two basic types of pili: **short attachment pili** and **long conjugation pili**.

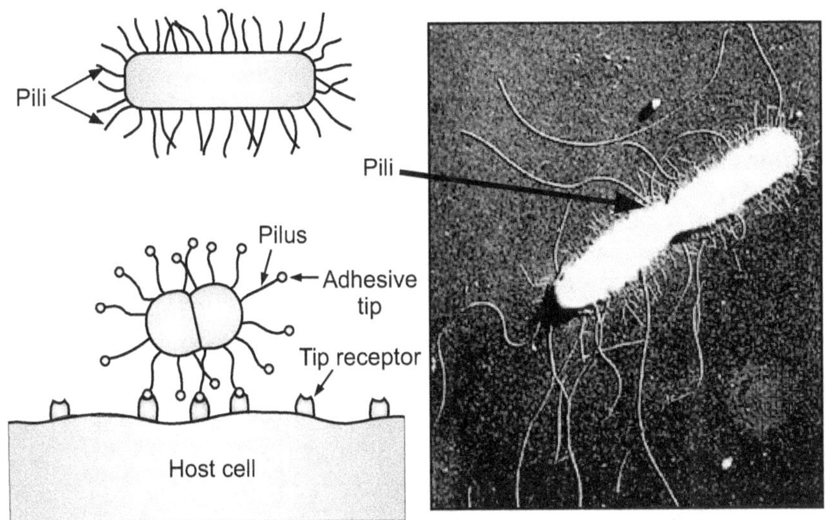

Fig. 1.20: Adhesive Tip of Bacterial Pili Binding to Host Cell Receptors

Salmonella showing both flagella and pili

Short attachment pili are usually short and quite numerous and enable bacteria to colonize environmental surfaces or cells and resist flushing.

Long conjugation pili, also called "F" or sex pili, that are longer and very few. The conjugation pilus enables conjugation. Conjugation is the transfer of DNA from one bacterium to another by cell-to-cell contact. In gram-negative bacteria it is typically the transfer of DNA from a donor or "male bacterium" with a sex pilus to a recipient or "female bacterium" to enable genetic recombination.

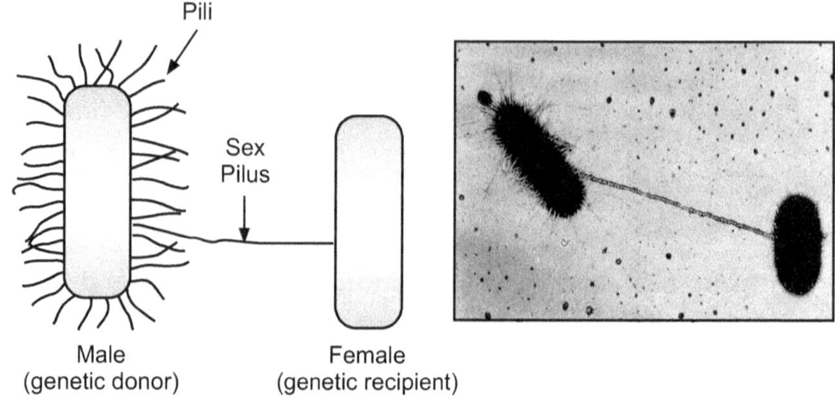

Fig. 1.21: Conjugation (Sex) Pilus *E. coli* bacteria exchanging genes

Proteins associated with bacterial fimbriae and pili function as antigens and initiate adaptive immunity.

1.3.9 Ribosomes

All prokaryotic and eukaryotic cells contain ribosomes. Ribosomes are centres of protein synthesis.

Prokaryotic ribosomes are slightly smaller than the ribosomes of eukaryotic cells and are seen integrated in linear strands of mRNA to form polysomes.

Ribosomes are composed of two subunits, each of which consists of protein and a type of RNA called ribosomal RNA (rRNA).

Prokaryotic ribosomes are smaller (70S) than eukaryotic ribosomes (80S).

Prokaryotic ribosomes consist of 30S and 50S, the smaller unit and the larger unit respectively whereas eukaryotic ribosomes have smaller subunit and larger subunit as 40S and 60S respectively.

In eukaryotes, rRNA in ribosomes has four strands whereas, in prokaryotes, rRNA is organized into three strands in ribosomes.

In eukaryotic cells, ribosomes are in free and bound form, whereas prokaryotic cells have only free form.

Eukaryotic cells have chloroplasts and mitochondria as organelles and those organelles also have ribosomes 70S. Therefore, eukaryotic cells have different types of ribosomes (70S and 80S), whereas prokaryotic cells only have 70S ribosomes.

Prokaryotic ribosomes differ from eukaryotic ribosomes in the number of proteins and rRNA molecules they contain. Eukaryotic ribosome consists of eight kinds of protein and five kinds of rRNA, and prokaryotic ribosomes are made of three types of rRNA and fifty kinds of protein.

The letter S refers to Svedberg units, which indicate the relative rate of sedimentation during ultra-high-speed centrifugation. Sedimentation rate is a function of the size, weight, and shape of a particle.

The subunits of a 70S ribosome are a small 30S subunit containing one molecule of rRNA and a larger 50S subunit containing two molecules of rRNA.

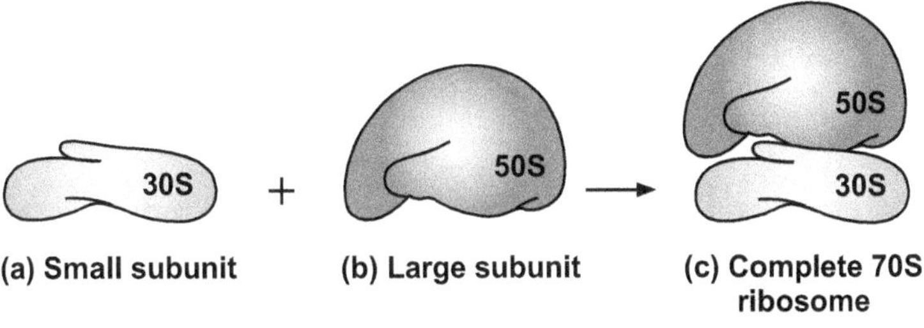

(a) Small subunit　　　(b) Large subunit　　　(c) Complete 70S ribosome

Fig. 1.22

Ribosomes are tiny spherical organelles that make proteins by joining amino acids together. Many ribosomes are found free in the cytosol, while others are attached to the rough endoplasmic reticulum.

The purpose of the ribosome is to translate messenger RNA (mRNA) to proteins with the help of tRNA.

In eukaryotes, ribosomes can commonly be found in the cytosol of a cell, the endoplasmic reticulum or mRNA, as well as the matrix of the mitochondria. Proteins synthesized in each of these locations serve a different role in the cell. In prokaryotes, ribosomes can be found in the cytosol of the cell.

This protein-synthesizing organelle is the only organelle found in both prokaryotes and eukaryotes. Ribosomes are not membrane bound.

Ribosomes are composed of two subunits, one large and one small, that only bind together during protein synthesis.

The purpose of the ribosome is to take the actual message and the charged aminoacyl-tRNA complex to generate the protein. To do so, they have three binding sites. One is for the mRNA; the other two are for the tRNA. The binding sites for tRNA are the A site, which holds the aminoacyl-tRNA complex, and the P site, which binds to the tRNA attached to the growing polypeptide chain.

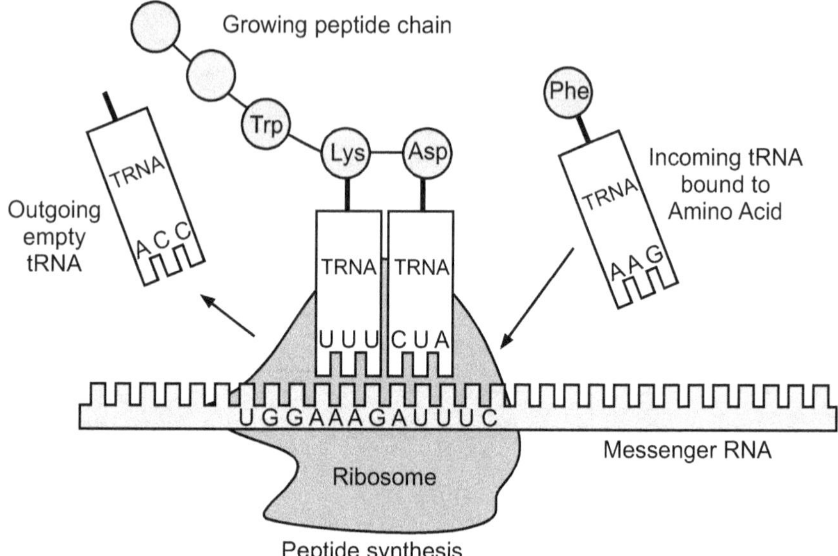

Fig. 1.23: Peptide synthesis by a ribosome

The ribosome assembles amino acids into a protein. The specific amino acids are controlled by the mRNA sequence.

The 70S ribosome is made up of a 50S and 30S subunits. The 50S subunit contains the 23S and 5S rRNA while the 30S subunit contains the 16S rRNA. These rRNA molecules differ in size in eukaryotes and are complexed with many ribosomal proteins, the number and type of which can vary slightly between organisms. The ribosome is the most commonly observed intracellular multiprotein complex in bacteria.

Ribosomes play a key role in the catalysis of two important and crucial biological processes. peptidyl transfer and peptidyl hydrolysis.

1.3.10 Chromosomal Material

The term genome refers to the sum of an organism's genetic material. The bacterial genome is composed of a single molecule of chromosomal deoxyribonucleic acid or DNA and is located in a region of the bacterial cytoplasm called the nucleoid. The bacterial DNA is packaged in loops back and forth. Unlike the eukaryotic nucleus, the bacterial nucleoid has no nuclear membrane. Since, bacteria are haploid, that is they have only one chromosome and only reproduce asexually, there is also no meiosis in bacteria.

The bacterial chromosome is one long, single molecule of double stranded, helical, supercoiled DNA. In most bacteria, the two ends of the double-stranded DNA covalently bond together to form both a physical and genetic circle. The chromosome is generally around 1000 μm long and frequently contains as many as 3500 genes. *E. coli*, a bacterium that is 2-3 μm in length, has a chromosome approximately 1400 μm long.

In general, DNA is replicated by uncoiling of the helix, strand separation by breaking of the hydrogen bonds between the complementary strands, and synthesis of two new strands by complementary base pairing. DNA topoisomerase enzymes are used to supercoil and relax the bacterial chromosome during DNA replication and transcription.

During DNA replication the nitrogenous base adenine forms hydrogen bonds with thymine and guanine forms hydrogen bonds with cytosine.

Genes located along the DNA are transcribed into RNA molecules, primarily messenger RNA (mRNA), transfer RNA (tRNA, and ribosomal RNA (rRNA). Messenger RNA is then translated into protein at the ribosomes. DNA determines what proteins and enzymes an organism can synthesize and, therefore, what chemical reactions it is able to carry out.

Plasmid:

The term plasmid was introduced in 1952 by the American molecular biologist Joshua Lederberg to refer to "any extrachromosomal hereditary material. A plasmid is a small DNA molecule within a cell that is physically separated from chromosomal DNA and can replicate independently. They are most commonly found as small circular, double-stranded DNA molecules in bacteria; however, plasmids are sometimes present in archaea and eukaryotic organisms. In nature, plasmids often carry genes that benefit the survival of the organism, such as by providing antibiotic resistance. While the chromosomes are big and contain all the essential genetic information for living under normal conditions, plasmids usually are very small and contain only additional genes that may be useful in certain situations or conditions. Artificial plasmids are widely used as vectors in molecular cloning, serving to drive the replication of recombinant DNA sequences within host organisms. In the laboratory, plasmids may be introduced into a cell via transformation.

Plasmids are considered replicons, units of DNA capable of replicating autonomously within a suitable host. Plasmids can be broadly classified into conjugative plasmids and non-conjugative plasmids. Conjugative plasmids contain a set of transfer or tra genes which promote sexual conjugation between different cells. Plasmids are transmitted from one bacterium to another (even of another species) mostly through conjugation. Non-conjugative plasmids are incapable of initiating conjugation, hence they can be transferred only with the assistance of conjugative plasmids. The size of the plasmid varies from 1 to over 200 kbp, and the number of identical plasmids in a single cell can range anywhere from one to thousands under some circumstances. A microbe can harbour different types of plasmids, but different plasmids can only exist in a single bacterial cell if they are compatible. If two plasmids are not compatible, one or the other will be rapidly lost from the cell. Different plasmids may therefore be assigned to different incompatibility groups depending on whether they can coexist together.

There are five main classes of plasmids based on the function:

1. **Fertility F-plasmids**, which contain tra genes. They are capable of conjugation and result in the expression of sex pili.

2. **Resistance (R) plasmids**, which contain genes that provide resistance against antibiotics or poisons. Historically known as R-factors, before the nature of plasmids was understood.

3. **Col plasmids**, which contain genes that code for bacteriocins, proteins that can kill other bacteria.

4. **Degradative plasmids**, which enable the digestion of unusual substances, e.g. toluene and salicylic acid.

5. Virulence plasmids, which turn the bacterium into a pathogen.

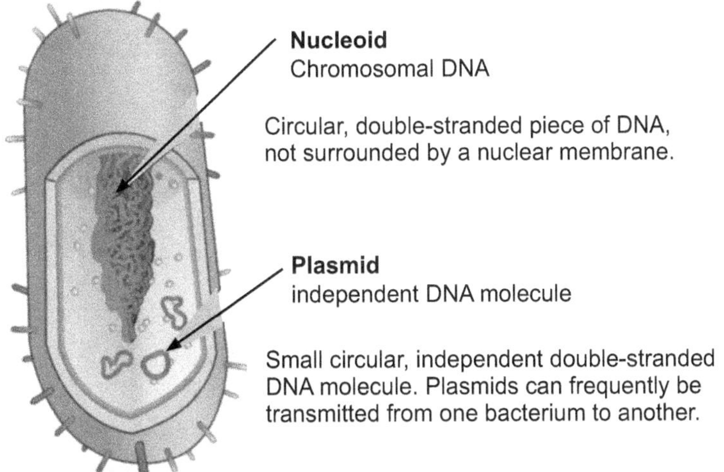

Fig. 1.24: Bacterial cell with chromosomal DNA and plasmids

Functions of Plasmids:

1. They may contain genes that enhance the survival of an organism, either by killing other organisms or by defending the host cell by producing toxins.

2. F-plasmids, contain transfer genes that allow genes to be transferred from one bacterium to another through conjugation.

3. Resistance or R plasmids contain genes that help a bacterial cell defend against environmental factors such as poisons or antibiotics.

4. When a virulence plasmid is inside a bacterium, it turns that bacterium into a pathogen, which is an agent of disease. The bacterium *Escherichia coli* has several virulence plasmids. Certain strains of *E. coli* can cause severe diarrhoea and vomiting.

5. Degradative plasmids help the host bacterium to digest compounds that are not commonly found in nature, such as camphor, xylene, toluene, and salicylic acid. These plasmids contain genes for special enzymes that break down specific compounds.

6. Col plasmids contain genes that make bacteriocins (also known as colicins), which are proteins that kill other bacteria and thus defend the host bacterium.

1.3.11 Cell Inclusions and Storage Granules

Within the cytoplasm of prokaryotic cells are several kinds of reserve deposits, known as inclusions. Cells may accumulate certain nutrients when they are abundant and use them when environment is deficient. Some inclusions are common to a wide variety of bacteria, whereas others are limited to a small number of species and therefore serve as basis for identification.

Most bacteria do not live in environments that contain large amounts of nutrients at all times. To accommodate these transient levels of nutrients, bacteria contain several different methods of nutrient storage that are employed in times of plenty, for use in times of want. For example, many bacteria store excess carbon in the form of poly-hydroxy-alkanoates or glycogen. Some microbes store soluble nutrients, such as nitrate in vacuoles. Sulphur is most often stored as elemental granules which can be deposited either intra- or extracellularly. Sulphur granules are especially common in bacteria that use hydrogen sulphide as an electron source.

The inclusion bodies are tiny particles found freely suspended and floating within the cytoplasmic matrix. Therefore, also referred to as cytoplasmic inclusions. These cell inclusions are formed with decreasing pH and from the pool of soluble fusion proteins within the cell. They are the elementary bodies, formed during infectious diseases or within the virus-infected cells such as rabies, herpes, measles, etc.

Inclusion bodies are non-living chemical compounds and by-products of cellular metabolism. They are found both in prokaryotes and eukaryotes. There are a wide variety of inclusion bodies in different types of cells. In prokaryotic cells, they are mainly formed to store reserve materials.

Common inclusion bodies in microorganisms:

Sr. No.	Inclusion bodies	Site	Composition	Function
1.	Gas vesicles	Aquatic bacteria Cyanobacteria	Protein shells Filled with gases	Buoyancy in vertical water column

Contd...

Sr. No.	Inclusion bodies	Site	Composition	Function
2.	Carboxysomes	Many autotrophic bacteria	Enzymes for autotrophic CO_2 Fixation	Site of CO_2 fixation
3.	PHB granules	Many bacteria e.g. Pseudomonas	Polymerised hydroxy butyrate	Reserve carbon and energy source
4.	Metachromatic granules	Many bacteria Corynebacterium	Linear polymers of PO_4	Reserve of high energy phosphates
5.	Glycogen bodies	Many bacteria e.g. *E. coli*	Polyglucose	Reserve C and energy source
6.	Starch granules	e.g. *E. coli*	Amylose and amylopectin	Reserve C source
7.	Magnetosomes	Certain aquatic bacteria	Magnetite Iron oxide Fe_3O_4	Orienting and migrating along geomagnetic field lines
8.	Sulphur granules	Phototrophic purple and green sulphur bacteria and lithotrophic colourless sulphur bacteria	Elemental sulphur	Reserve of electrons in phototrophs and reserve of energy source in lithotrophs
9.	Chlorosomes	Green bacteria	Bacterio-chlorophyll, proteins and lipids	Light harvesting pigments and antennae

1. Gas Vesicles:

Hollow cavities found in many aquatic prokaryotes, cyanobacteria, anoxygenic photosynthetic bacteria, and halobacteria are called gas vacuoles. Each vacuole consists of rows of several individual gas vesicles, which are hollow cylinders covered by protein. Gas vesicles are aggregates of many hollow structures containing various gases. These are membrane bound structures permeable to gases. Gas vesicles maintain buoyancy and allows floating property to bacteria so that the cells can remain at the depth in the water appropriate for them to receive enough amounts of oxygen, light, and nutrients. They make the cells lighter and are observed under light microscope as dense refractile bodies.

Gas vesicles occur in five phyla of the Bacteria and two groups of the Archaea, but they are mostly restricted to planktonic microorganisms, in which they provide buoyancy.

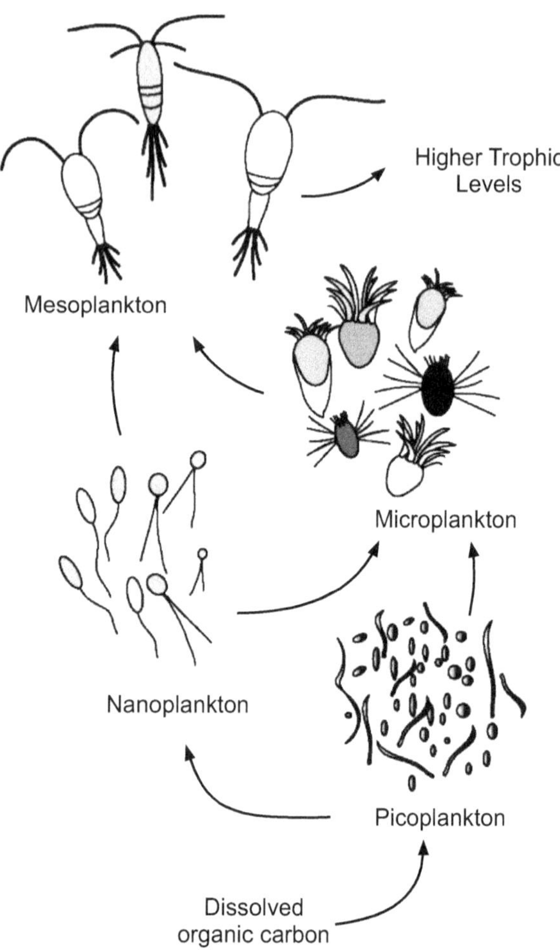

Fig. 1.25: Gas vesicles provide buoyancy for some planktonic bacteria by decreasing their overall cell density

Gas vesicles are spindle-shaped structures found in some planktonic bacteria that provides buoyancy to these cells by decreasing their overall cell density. Positive buoyancy is needed to keep the cells in the upper reaches of the water column, so that they can continue to perform photosynthesis. They are made up of a shell of protein that has a highly hydrophobic inner surface, making it impermeable to water but permeable to most gases. Because the gas vesicle is a hollow cylinder, it is liable to collapse when the surrounding pressure becomes too great.

Gas vesicles provide bouyancy for planktonic bacteria by decreasing their overall cell density. There is a simple relationship between the diameter of the gas vesicle and pressure at which it will collapse – the wider the gas vesicle the weaker it becomes. However, wider gas vesicles are more efficient. They provide more buoyancy per unit of protein than narrow gas vesicles. Different species produce gas vesicles of different diameters, allowing them to colonize different depths of the water column fast growing, highly competitive species with wide gas vesicles in the topmost layers; slow growing, dark-adapted, species with strong narrow gas vesicles in the deeper layers. Gas vesicles are found in *Enterobacterium, Serratia, cyanobacteria, heterotrophic bacteria*, green sulphur bacteria like *Thiospirillum, Thiocystis, Chromatium* and *Archaea*.

2. Carboxysomes

Carboxysomes were first seen in 1956, in the cyanobacterium *Phormidium uncinatum*. They were first purified from *Thiobacillus neapolitanus* in 1973 and were shown to contain ribulose 1,5-diphosphate carboxylase (RuBisCo) held within a rigid outer covering. Photosynthetic bacteria use carbon dioxide as their sole source of carbon and require ribulose 1,5-diphosphate carboxylase enzyme for carbon dioxide fixation which is achieved by Calvin cycle. Among the bacteria containing carboxysomes are Nitrifying bacteria like *Nitrobacter, Nitrococcus*, autotrophic bacteria like *Cyanobacteria, Thiobacillus* and many chemotrophic bacteria that fix carbon dioxide.

They are proteinaceous polyhedral bodies with 80 - 140 nm diameter resembling phage heads in their morphology. These are bounded by single layered membrane. *Carboxysomes* harbour the B_{12}-containing coenzyme glycerol dehydratase, the key enzyme of glycerol fermentation to 1,3-propanediol, in some Enterobacteriaceae, such as *Salmonella*.

3. Poly-β-hydroxybutyric acid (PHB)granules

A common lipid storage material is the polymer poly-β-hydroxybutyric acid (PHB) granules. PHB granules are non-nitrogenous organic storage granules found in bacteria *Azotobacter, Bacillus cereus, Bacillus megaterium, Corynebacterium, Pseudomonas, Rhodospirillum, Spirillum,* Cupriavidus *necator, Methylobacterium rhodesianum* etc. PHB was first isolated and characterized in 1925 by French scientist Maurice Lemoigne. These granules are deposited evenly in cytoplasm and are detectable under light microscope using Sudan black stain. They appear as refractile bodies of variable size. They can be extracted by organic solvents. They are formed in response to conditions of physiological stress like nitrogen or sometimes oxygen starvation and high availability of carbon. The polymer is primarily a product of carbon assimilation (from glucose or starch) and is employed by microorganisms as a form of energy storage molecule to be metabolized when other common energy sources are not available. Microbial biosynthesis of PHB starts with the condensation of two molecules of acetyl-CoA to give acetoacetyl-CoA which is subsequently reduced to hydroxy butyryl-CoA. This later compound is then used as a monomer to polymerize PHB. So PHB is primarily a product of carbon assimilation and is used by micro-organisms as a form of energy storage molecules. They are thermoplastic polymers and are totally biodegradable.

4. Metachromatic Granules:

These granules are found in many organisms such as algae, fungi, bacteria and as well as protozoans. Metachromatic granules are large inclusion which stores inorganic phosphate that can be used in the synthesis of ATP. In these granules linear chains of orthophosphates are joined to each other by ester bonds. They are also known as volutin granules. They are basophilic in nature and therefore stained by basic dyes. They show metachromatic effect i.e. when these granules are stained with blue coloured basic dyes such as methylene blue, crystal violet, they appear red or brown due to interaction between dye and granules. It is generally formed by cells that grow in phosphate rich environments. They are generally formed during starvation of sulphur. *Corynebacterium diphtheriae* and *Lactobacillus* are a common example of bacteria possessing metachromatic granules. They are also found in the cytoplasm of *Saccharomyces*.

5. Glycogen Bodies:

These are granules containing glycogen i.e. polymer containing α 1-4 glycosidic linkages. They are distributed throughout the cytoplasm in high number. They are observed reddish brown by staining with iodine. They are of size 20 - 100 nm and are formed under limited nitrogen supply but with ample carbon supply. They are found in *Proteus, Klebsiella, Shigella, Bacillus, Salmonella,* and in *Cyanobacteria*. The limited nitrogen supply lowers nucleic acid and protein synthesis which accumulates excess carbon. This carbon is stored in the form of glycogen bodies. For re-utilization of carbon glycogen is degraded by phosphorylase enzyme.

6. Starch Granules:

Starch granules are the most important energy reserve in higher plants. They are composed principally of amylopectin (major fraction) and amylose (minor fraction). Their structure is assumed to consist of concentric shells of alternating hard, semi-crystalline, and soft amorphous layers. Production of insoluble Starch-Like Granules in *Escherichia coli* takes place by Modification of the Glycogen Synthesis Pathway. Starch granules are used as reserve food by the microorganisms

7. Magnetosomes:

Magnetosomes are inclusions of iron oxide (Fe_3O_4), formed by several gram-negative bacteria such as *Magnetospirillium magnetotacticum*, that act like magnets. Bacteria may use magnetosomes to move downward until they reach a suitable attachment site. *In-vitro*, magnetosomes can decompose hydrogen peroxide, which forms in cells in the presence of oxygen. Therefore, magnetosomes may protect the cell against hydrogen peroxide accumulation.

Magnetosomes are intracellular organelles found in magnetotactic bacteria that allow them to sense and align themselves along a magnetic field (magnetotaxis). They contain 15 to 20 magnetite crystals that together act like a compass needle to orient magnetotactic bacteria in geomagnetic fields, thereby simplifying their search for their preferred microaerophilic environments. Each magnetite crystal within a magnetosome is surrounded by a lipid bilayer. Specific soluble and transmembrane proteins are sorted to the membrane. Recent research has shown that magnetosomes are invaginations of the inner membrane. Magnetite-bearing magnetosomes have also been found in eukaryotic magnetotactic algae, with each cell containing several thousand crystals.

Magnetotactic bacteria usually mineralize either iron oxide magnetosomes, which contain crystals of magnetite (Fe_3O_4), or iron sulphide magnetosomes, which contain crystals of greigite (Fe_3S_4).

The morphology of magnetosome crystals varies, but is consistent within cells of a single magnetotactic bacterial species. They may be roughly cuboidal, elongated prismatic (roughly rectangular), and tooth-, bullet-, or arrowhead-shaped. Magnetosome crystals are typically 35–120 nm long. In most magnetotactic bacteria, the magnetosomes are arranged in one or more chains.

Magnetic interactions between the magnetosome crystals in a chain cause their magnetic dipole moments to orientate parallel to each other along the length of the chain. The magnetic dipole moment of the cell is usually large enough so that its interaction with Earth's magnetic field overcomes thermal forces that tend to randomize the orientation of the cell in its aqueous surroundings. Magnetotactic bacteria also use aerotaxis, a response to changes in oxygen concentration that favours swimming toward a zone of optimal oxygen concentration. In lakes or oceans, the oxygen concentration is commonly dependent on depth. If the Earth's magnetic field has a significant downward slant, the orientation along field lines aids the search for the optimal concentration. This process is called magneto-aerotaxis.

8. Sulphur Granules:

The "sulphur bacteria" that belong to the genus *Thiobacillus* – derive energy by oxidizing sulphur – containing compounds. These bacteria may deposit sulphur granules in the cell, where they serve as an energy reserve.

9. Chlorosomes:

Chlorosomes were discovered in 1964 and described as rectangle bodies attached to the inner side of the cytoplasmic membrane in thin sections of cells from *Chlorobium* species. A chlorosome is a photosynthetic light-harvesting complex found in anoxygenic green bacteria, green sulphur bacteria (GSB) and some green filamentous anoxygenic phototrophs. Chlorosomes are flattened ellipsoidal organelles appressed to the cytoplasmic face of the cell membrane. They are composed of bacteriochlorophylls with some proteins, lipids, carotenoids and quinones. Proteins are confined to the surface of the chlorosome while most bacteriochlorophyll molecules are found within the interior where they assemble into aggregates. Green sulfur bacteria are a group of organisms that generally live in extremely low-light environments, such as at depths of 100 metres in the Black Sea. The ability to capture light energy and rapidly deliver it to where it needs to go is essential to these bacteria, some of which see only a few photons of light per chlorophyll per day. To achieve this, the bacteria contain chlorosome structures, which contain up to 250,000 chlorophyll molecules. Chlorosomes are ellipsoidal bodies, in Green Sulphur Bacteria their length varies from 100 to 200 nm, width of 50 - 100 nm and height of 15 - 30 nm.

A unique property of chlorosomes is that their main pigments, bacteriochlorophylls (BChl) c, d or e, are organized in the form of an aggregate, in contrast to other photosynthetic light-harvesting complexes where proteins maintain the distances and mutual orientations between pigments. The aggregates inside a chlorosome are composed of many thousands of tightly associated BChl molecules making the chlorosome the largest light-harvesting complex known to date. The aggregation also leads to strong excitonic coupling between the BChls which provides the basis for high light-harvesting efficiency. Consequently, some of the chlorosome-possessing bacteria can grow under extremely low light conditions and chlorosomes are considered the most efficient antenna found in nature.

Bacteria which contain chlorosomes include:

- *Chlorobium limicola*
- *Chlorobium phaeobacteroides*
- *Chlorobium vibrioforme*
- *Chlorobium tepidum*
- *Pelodictyon lutoleum*
- *Chloroflexus aurantiacus*
- *Chloroflexus aggregans*
- *Chloronem agiganteum*
- *Oscillochloris trichoides*
- *Chloracidobacterium thermophilum*

Think Over It

1. How is bacterial cell different from human cell?
2. What is the microbiome?
3. How are bacterial endospores structurally and chemically different from the vegetative cells that form them?
4. How do the bacteria move?
5. What is a bacterial biofilm?

SUMMARY

- Bacteria are 0.2 - 2.0 µm in diameter and 2 - 8 µm in length.
- The three basic bacterial shapes are coccus (spherical), bacillus (rod shaped) and spiral (twisted).

- The glycocalyx i.e. capsule and slime later contains gelatinous polysaccharide. Capsule protects the pathogens from phagocytosis.
- Flagella are longer filamentous appendages consisting of filament, hook and basal body.
- Fimbriae and pili are short and thin appendages.
- The cell wall surrounds the plasma membrane and protects the cell from changes in water pressure.
- The cell wall consists of peptidoglycan, a polymer consisting of NAG and NAM and short chains of amino acids.
- Mycoplasma is a bacterial genus that naturally lacks cell walls.
- The plasma membrane encloses the cytoplasm and is a lipid bilayer with peripheral and integral proteins (fluid mosaic model).
- The cytoplasm is the fluid component inside the plasma membrane.
- The nucleus contains the DNA of the bacterial Chromosome.
- The cytoplasm contains numerous 70S ribosomes which consists of rRNA and proteins.
- Endospores are resting structures which allow survival of bacteria during adverse environmental conditions.
- Cell inclusions are reserve deposits found in bacteria.

Exercise

(A) Questions for One Mark:

1. What is the typical size of a bacterium?
2. Why are different shapes of bacteria?
3. What accounts for the rigidity and strength of bacterial cell walls?
4. What are carboxysomes?
5. What is spheroplast?
6. What is protoplast?

7. What are L forms?

8. What is the difference between bacterial chromosome and plasmid DNA?

9. Component responsible for bacterial endospore resistance is,
   a) Ca-dipicolinate
   b) Na-dipicolinate
   c) Colinic acid
   d) All of the above.

10. Volutin granules are also called as,
    a) Babe's Granules
    b) PHB Granules
    c) Volatile Granule
    d) None of the above.

11. A "hair" like structure involved in chemotactic response of bacterium is called as
    a) Flagella
    b) Pili
    c) fimbriae
    d) All of the above

(B) Questions for Two Marks:

1. What arrangements of flagella are found on bacteria?

2. What are chemotaxis, phototaxis, aerotaxis, and magnetotaxis?

3. What is the significance of the cell wall lipopolysaccharide of Gram-negative bacteria?

4. What is the function of the chemical nature and function of the following cell inclusions in bacterial cells? (each for 2 marks)
   (i) Glycogen bodies
   (ii) PHB granules
   (iii) Metachromatic granules
   (iv) Carboxysomes
   (v) Magnetosomes

5. Write any two examples of
   (i) Encapsulated bacteria
   (ii) Bacteria having spore
   (iii) Gram positive bacteria

    (iv) Gram negative bacteria

    (v) Bacteria containing metachromatic granules

    (vi) Bacteria containing PHB granules

    (vii) Bacteria having flagella

6. How does a protoplast differ from a spheroplast?

7. Under what conditions are endospores formed by bacteria?

8. Diagrammatically represent various shapes and arrangement of bacteria.

9. Describe the functions of flagella

(C) Questions for Four Marks:

1. Illustrate diagrammatically

    (i)   Endospore

    (ii)   Plasma membrane

    (iii) Gram positive cell wall

    (iv) Gram negative cell wall

2. What is the location, composition and functions (any two) of each of the following bacterial structures (each for 4 marks)

    (i)   Flagella

    (ii)   Endospore

    (iii) Fimbriae or Pili

    (iv) Capsule

    (v)   Plasma membrane

    (vi) Cell wall

    (vii) Ribosomes

    (viii) Chromosome

    (ix) Plasmid

3. What are the similarities and differences between bacterial fimbriae and flagella?

4. Differentiate between Gram positive and Gram-negative cell wall.

5. Elaborate on structure of Gram-positive cell wall.

6. Elaborate on structure of Gram negative cell wall.

7. Describe the functions carried out by cell membrane.

8. Describe the structure of cell wall with respect to Fluid mosaic model.

9. Explain the stages in endospore formation.

10. Elaborate on spore germination

11. Elaborate on Gram positive and Gram-negative flagella.

12. Explain the Mechanism of Flagellar motility.

13. What is the role of ribosomes?

14. What are the functions carried out by the plasmids?

15. Describe magnetosomes and gas vesicles.

16. Describe the structure and chemical composition of gram positive bacterial cell wall.

17. Describe the structure and chemical composition of gram negative bacterial cell wall.

(D) Write short notes on (Four marks each)

1. Gas vesicles

2. Carboxysomes

3. PHB granules

4. Metachromatic granules

5. Glycogen bodies

6. Magnetosomes

7. Fluid mosaic model of cell membrane

8. Chlorosomes

9. Peptidoglycan

10. Spore germination

11. Functions of capsule

12. Ribosomes

Chapter 2...

Chemical Basis of Microbiology

Learning Objectives...

➢ To understand the basic molecular structure and functions of biomolecules which make up the chemical basis of microbiology.

➢ To understand the classification of bacteria and viruses.

Francis Harry Compton Crick (8 June, 1916 – 28 July 2004) was a British molecular biologist, biophysicist, and neuroscientist. In 1953, he co-authored with James Watson the academic paper proposing the double helix structure of the DNA molecule. Together with Watson and Maurice Wilkins, he was jointly awarded in 1962 Nobel Prize in "Physiology or Medicine" for their discoveries concerning the molecular structure of nucleic acids and its significance for information transfer in living material.

Francis Harry Compton Crick

Crick was an important theoretical molecular biologist and played a crucial role in research related to revealing the helical structure of DNA. He is widely known for the use of the term "Central dogma" to summarize the idea that once information is transferred from nucleic acids (DNA or RNA) to proteins, it cannot flow back to nucleic acids. In other words, the final step in the flow of information from nucleic acids to proteins is irreversible.

2.1 ATOM

An atom is a fundamental piece of matter that uniquely defines a chemical element. An atom is made up of three kinds of particles called subatomic particles: protons, neutrons, and electrons. The protons and the neutrons make up the centre of the atom called the Nucleus. One or more electrons usually surround the nucleus. The nucleus is positively charged.

2.2 BIOMOLECULES

Biomolecules are the building blocks of life. They perform important functions in living organisms. Biomolecules are essential for typical biological processes, such as cell division, morphogenesis, or development. Large macromolecules such as amino acids, lipids, carbohydrates, proteins, polysaccharides, and nucleic acids as well as small molecules such as primary metabolites, secondary metabolites, and natural products are included in biomolecules.

Biomolecules are produced within the organisms i.e. usually endogenous. But organisms depend upon exogenous biomolecules, to survive, for example: certain nutrients.

2.3 TYPES OF BONDS

Covalent: This is a chemical bond that involves the sharing of electron pairs between two atoms. These electron pairs are known as bonding electron pairs, and they share these electrons to form covalent bond. This bonding is primarily found between non-metals; however, non-metals and metals also show covalent bonding.

Non-covalent: This is a type of chemical bond that is formed between macromolecules. They do not involve sharing a pair of electrons. Non-covalent bonds are used to bond large molecules such as proteins and nucleic acids. Non-covalent interactions involve different mechanisms. They include van der Waal's interactions, hydrogen bonding, and electrostatic interactions (called ionic bonding).

2.4 LINKAGES

Ester: An ester bond is the bond between an alcohol group (–OH) and a carboxylic acid group (–COOH), formed when a molecule of water (H_2O) is eliminated. Fats are esters, produced by the bonding of fatty acids with the alcohol glycerol. Ethyl acetate is an ester. The hydrogen on the carboxyl group of acetic acid is replaced with an ethyl group. Other examples of esters include ethyl propanoate, propyl methanoate, propyl ethanoate, and methyl butanoate. Glycerides are fatty acid esters of glycerol.

Phosphodiester: This bond occurs when exactly two of the hydroxyl groups in phosphoric acid react with hydroxyl groups on other molecules to form two ester bonds. Each ester bond is formed by a condensation reaction in which water is lost. The phosphodiester bond, links the sugar molecules and phosphate molecules in the backbone for building DNA and RNA. This is one of the most crucial components as it maintains the integrity of the genetic code.

Peptide Bond: This is a chemical bond formed between two molecules when the carboxyl group of one molecule reacts with the amino group of the other molecule, releasing a molecule of water (H_2O). This is a dehydration synthesis reaction. The strength of the peptide bond largely depends on the resonance between nitrogen and the carbonyl group. The peptide bond is rigid, planar, and stronger than a typical C-N single bond condensation reaction, and usually occurs between amino acids.

Glycosidic Bond: A glycosidic bond or glycosidic linkage is a type of covalent bond that joins a carbohydrate (sugar) molecule to another group, which may or may not be another carbohydrate. They form by a condensation reaction between an alcohol or amine of one molecule and the anomeric carbon of the sugar and, therefore, may be O-linked or N-linked.

An aldehyde or a ketone group on the sugar can react with a hydroxyl group on another sugar; forming glycosidic bond.

There are two types of glycosidic bonds : 1, 4 alpha and 1, 4 beta glycosidic bonds. 1, 4 alpha glycosidic bonds are formed when the OH on the carbon-1 is below the glucose ring; while 1, 4 beta glycosidic bonds are formed when the OH is above the plane.

2.5 CARBOHYDRATES

The term carbohydrate is a combination of the "hydrates of carbon". They are also known as "Saccharides" from the Greek word "Sakcharon" which means sugar.

Carbohydrates are compounds made from three elements: carbon, hydrogen and oxygen. Monosaccharides (e.g. glucose) and disaccharides (e.g. sucrose) are relatively small molecules. They are often called sugars. Other carbohydrate molecules are very large (polysaccharides such as starch and cellulose).

Carbohydrates are:

- A source of energy for the body e.g. glucose and a store of energy, e.g. starch in plants.

- Building blocks for polysaccharides (giant carbohydrates), e.g. cellulose in plants and glycogen in the human body.
- Components of other molecules e.g. DNA, RNA, glycolipids, glycoproteins and ATP.

The main classification of carbohydrate is done on the basis of hydrolysis. This classification is as follows:

1. **Monosaccharides:** These are the simplest form of carbohydrate that cannot be hydrolyzed further. They have the general formula of $(CH_2O)_n$. E.g. glucose, ribose etc.

2. **Oligosaccharides:** The carbohydrates which on hydrolysis yield two to ten smaller units or monosaccharides are oligosaccharides.

3. **Disaccharides:** These give two units of the same or different monosaccharides on hydrolysis. For example, sucrose on hydrolysis gives one molecule of glucose and fructose each. Maltose on hydrolysis gives two molecules of only glucose.

4. **Trisaccharides:** These on hydrolysis give three molecules of monosaccharides, same or different. E.g. Raffinose.

5. **Tetrasaccharides:** These on hydrolysis give four molecules of monosaccharides. E.g. Stachyose.

6. **Polysaccharides:** These give a large number of monosaccharides when they undergo hydrolysis; these carbohydrates are not sweet in taste and are also known as non-sugars. E.g. starch, glycogen etc.

2.5.1 Monosaccharides

Monosaccharides are the simplest carbohydrates which are often called sugars. They are the building blocks from which all bigger carbohydrates are made.

Monosaccharides have the general molecular formula $(CH_2O)_n$, where n can be 3, 5 or 6. They can be classified according to the number of carbon atoms in a molecule:

n = 3	trioses, e.g. glyceraldehyde
n = 5	pentoses, e.g. ribose and deoxyribose ('pent' indicates 5)
n = 6	hexoses, e.g. fructose, glucose and galactose ('hex' indicates 6)

Monosaccharides containing the aldehyde group are classified as aldoses, and those with a ketone group are classified as ketoses. Aldoses are reducing sugars; ketoses are non-reducing sugars. This is important in understanding the reaction of sugars. However, in water pentoses and hexoses exist mainly in the cyclic form, and in this form that they combine to form larger saccharide molecules.

Ribose is a carbohydrate with the formula $C_5H_{10}O_5$.

It is pentose monosaccharide (simple sugar) with linear form H–(C=O)–(CHOH)$_4$–H, which has all the hydroxyl groups on the same side in the Fischer projection. The term usually indicates d-ribose, which occurs widely in nature. Its synthetic mirror image, l-ribose, is not found in nature. Ribose is an aldopentose (a monosaccharide containing five carbon atoms) that, in its open chain form, has an aldehyde functional group at one end.

Deoxyribose, is a monosaccharide with idealized formula H–(C=O) – (CH$_2$) – (CHOH)$_3$–H. It is a deoxy sugar. It is derived from the sugar ribose by loss of an oxygen atom. 2-deoxyribose is a precursor to the nucleic acid DNA. 2-deoxyribose is an aldopentose, a monosaccharide with five carbon atoms and having an aldehyde functional group.

As a component of DNA, 2-deoxyribose derivatives have an important role in biology. The DNA (deoxyribonucleic acid) molecule, consists of a long chain of deoxyribose containing units called Nucleotides, linked via phosphate groups.

Glucose is a monosaccharide with formula $C_6H_{12}O_6$ or H-(C=O)-(CHOH)$_5$-H, whose five hydroxyl (OH) groups are arranged in a specific way along its six-carbon back. Glucose is usually present in solid form as a monohydrate with a closed pyran ring (dextrose hydrate). Glucose is a monosaccharide containing six carbon atoms and an aldehyde group and hence is an aldohexose. The glucose molecule can exist in an open chain (acyclic) and ring (cyclic) form.

Glucose is the most abundant monosaccharide. When solubilised in water or in the solid form, d-(+)-glucose is dextrorotatory, meaning it will rotate the direction of polarized light clockwise.

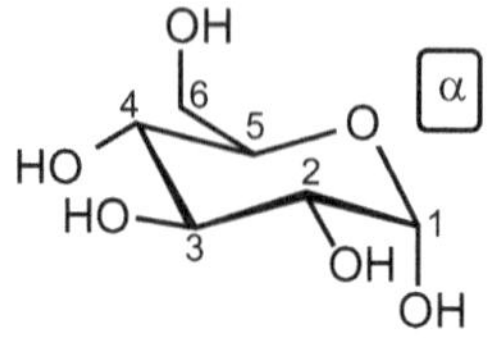

α-d-glucopyranose (chair form)

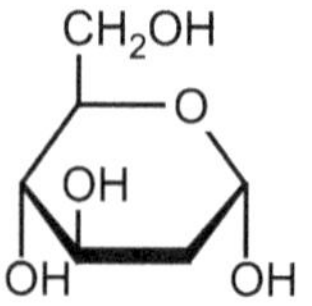

Haworth projection of α-d-glucopyranose

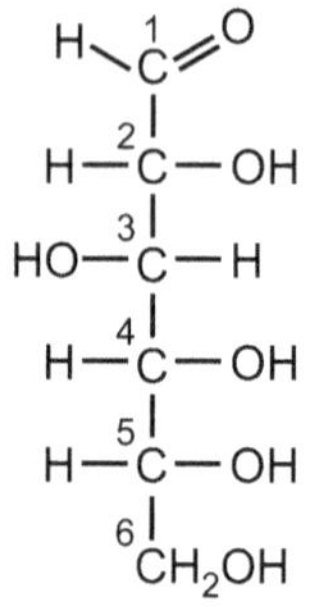

Fischer projection of d-glucose

Galactose is a monosaccharide and has the same chemical formula as glucose, i.e., $C_6H_{12}O_6$. It is similar to glucose in its structure, differing only in the position of one hydroxyl group. This difference, however, gives galactose different chemical and biochemical properties than glucose. Galactose is an energy-providing nutrient and a necessary basic substrate for the biosynthesis of many macromolecules in the body. Galactose is an important constituent of the complex polysaccharides, Galactose exists in both open-chain and cyclic form. The open-chain form has a carbonyl at the end of the chain. Four isomers are cyclic, two of them with a pyranose (six-membered) ring, two with a furanose (five-membered) ring.

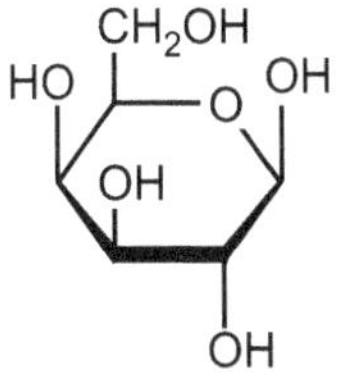

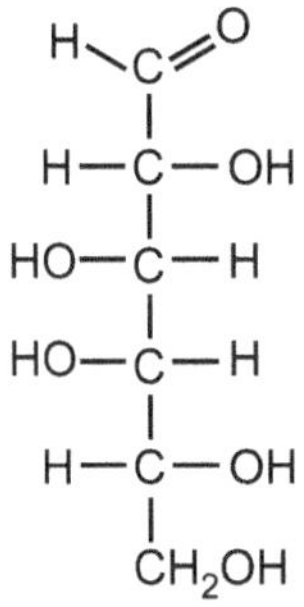

Fructose is a simple ketonic monosaccharide. Fructose is a 6-carbon polyhydroxyketone. Crystalline fructose adopts a cyclic six-membered structure owing to the stability of its hemiketal and internal hydrogen bonding. This form is formally called d-fructopyranose. In water solution, fructose exists as an equilibrium mixture of 70% fructopyranose and about 22% fructofuranose, as well as small amounts of three other forms, including the acyclic structure.

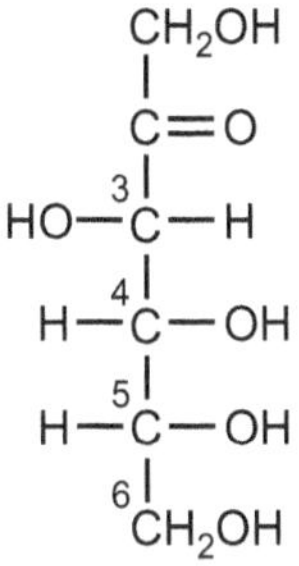

D-Fructose 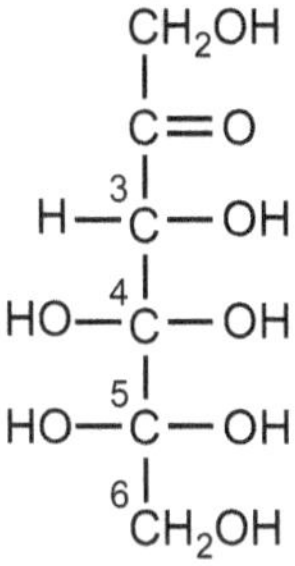   **L-Fructose**

Fructose has higher water solubility than other sugars, as well as other sugar alcohols. Fructose is, therefore, difficult to crystallize from an aqueous solution.

Like glucose, fructose is a source of energy for the cells. Cells process fructose to extract energy through a process called Aerobic respiration, which essentially means burning of fructose in the presence of oxygen to produce ATP, the cellular energy molecule.

2.5.2 Disaccharides

Disaccharides are sugars (carbohydrate molecules) that form when two simple sugars i.e. monosaccharides combine to form a disaccharide. Cyclic monosaccharides react with alcohols to form acetals and ketals. So when this happens individual monosaccharides linked together to form an acetal. This linkage is known as *glycosidic linkage.*

This linkage is an oxide linkage formed by the loss of a water molecule. When two monosaccharides linked together by glycosidic linkage, the resulting product is a disaccharide.

- The molecular formula of sucrose is $C_{12}H_{22}O_{11}$.

- If sucrose goes through acid catalysed hydrolysis it will give one mole of D-glucose and one mole of D-fructose.

- The chemical structure of sucrose comprises of α-form of glucose and β-form of fructose.

- The glycosidic linkage is α-linkage because the molecule formation is in α orientation

- Sucrose is a non-reducing sugar. As you can see from the structure it is combined (linked) at the hemiacetal oxygen and does not have a free hemiacetal hydroxide

- Since, it has no free hemiacetal hydroxide and it does not show mutarotation (α to β conversion). Sucrose also does not form osazones for the same reason.

- We can prove the structural formula of sucrose by hydrolysing with α-glycosidase enzymes which only hydrolyses alpha glucose. This test is positive for sucrose.

Sucrose

Lactose

Lactose is the primary ingredient found in the milk of all mammals. Unlike the majority of saccharides, lactose is not sweet to taste. Lactose consists of one galactose carbohydrate and one glucose carbohydrate. These are bound together by a 1-4 glycosidic bond in a beta orientation.

The structure of glucose and galactose differs. Galactose's fourth carbon has a different orientation in lactose than in sucrose. If it were not so the resulting molecule would have just been sucrose (glucose + glucose) instead of lactose.

Also from the structure, we can notice that lactose is a reacting sugar since it has one free hemiacetal hydroxide. Therefore, when we react lactose with bromine water it will give monocarboxylic acid.

2.5.3 Polysaccharides

Polysaccharides are polymeric carbohydrate molecules composed of long chains of monosaccharide units bound together by glycosidic linkages. On hydrolysis they give the constituent monosaccharides or oligosaccharides. They range in structure from linear to highly branched. Examples include storage polysaccharides such as starch and glycogen, and structural polysaccharides such as cellulose and chitin.

When all the monosaccharides in a polysaccharide are of the same type, the polysaccharide is called as a *homopolysaccharide* or *homoglycan*, but when more than one type of monosaccharide is present, they are called as *heteropolysaccharides* or *heteroglycans*.

Polysaccharides contain more than ten monosaccharide units, whereas oligosaccharides contain three to ten monosaccharide units.

Polysaccharides are an important class of biological polymers. Their function in living organisms is usually either structure or storage related. Starch (a polymer of glucose) is used as a storage polysaccharide in plants, being found in the form of both amylose and the branched amylopectin. In animals, the structurally similar glucose polymer is the more densely branched glycogen. The properties of glycogen allows it to be metabolized more quickly, for the active lives of moving animals.

Cellulose and chitin are examples of structural polysaccharides. Cellulose is present in the cell walls of plants and other organisms. It has many uses such as a significant role in the paper and textile industries, cellulose acetate, celluloid, and nitrocellulose. Chitin has a similar structure, but has nitrogen-containing side branches, increasing its strength. It is found in arthropod exoskeletons and in the cell walls of some fungi. It also has multiple uses, including surgical threads. Polysaccharides also include callose or laminarin, chrysolaminarin, xylan, arabinoxylan, mannan, fucoidan and galactomannan.

Starch:

Starch is often produced in plants as a way of storing energy. It exists in two forms: **amylose** and **amylopectin**. Both are made from α-glucose. Amylose is an unbranched polymer of α-glucose. The molecules coil into a helical structure. It forms a colloidal suspension in hot water. Amylopectin is a branched polymer of α-glucose. It is completely insoluble in water.

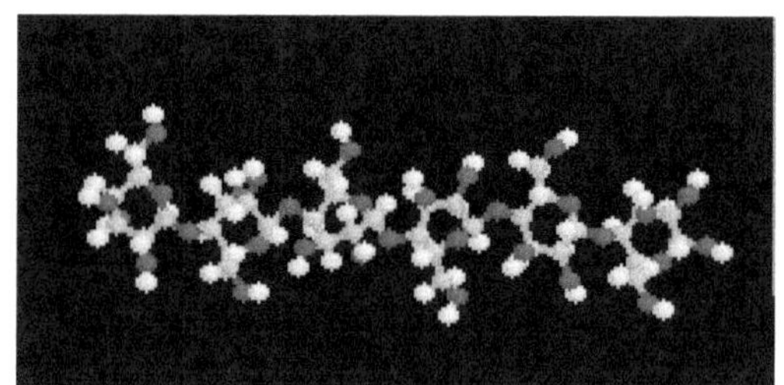

(a)

(b) A section of the amylose molecule

(c)

Fig. 2.1: Section of the amylopectin molecule

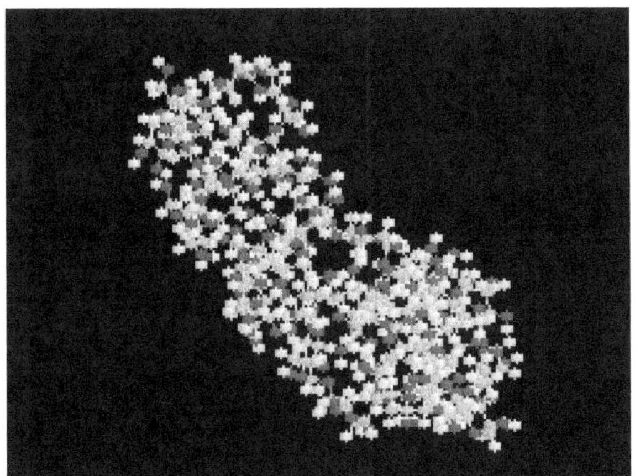

Fig. 2.2: A section of the amylopectin molecule

Glycogen:

Glycogen is amylopectin with very short distances between the branching side-chains. Inside the cell, glucose can be polymerised to make glycogen, which acts as a carbohydrate energy store. Glycogen is a polymer of glucose (up to 120,000 glucose residues) and is a primary carbohydrate storage form in animals. The polymer is composed of units of glucose-linked alpha (1-4) with branches occurring alpha (1-6) approximately every 8-12 residues. Due to the way glycogen is synthesised, every glycogen granule has at its core a glycogenin protein.

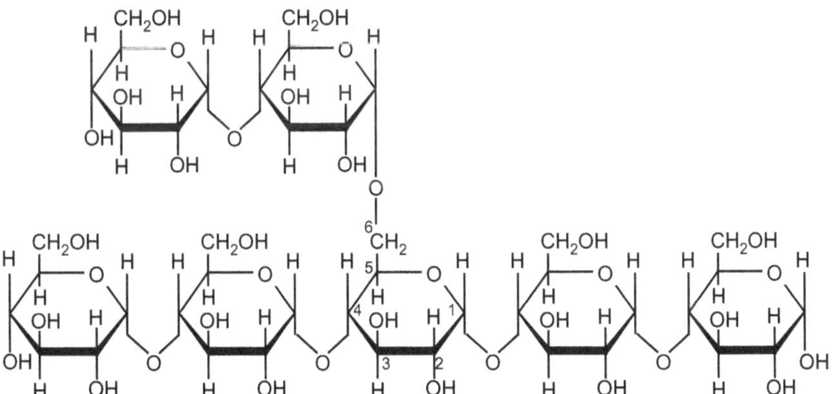

Fig. 2.3: Structure of Glycogen molecule

The glycogen branching enzyme catalyzes the transfer of a terminal fragment of six or seven glucose residues from a non-reducing end to the C-6 hydroxyl group of a glucose residue deeper into the interior of the glycogen molecule. The branching enzyme can act upon only a branch having at least 11 residues, and the enzyme may transfer to the same glucose chain or adjacent glucose chains.

Peptidoglycan, also known as **murein**, is a polymer consisting of sugars and amino acids that forms a layer outside the plasma membrane of bacteria forming the cell wall. The sugar component consists of alternating residues of β-(1, 4) linked N-acetylglucosamine and N-acetylmuramic acid. Attached to the N-acetylmuramic acid is a peptide chain of three to five amino acids. The peptide chain can be cross-linked to the peptide chain of another strand forming the 3D mesh-like layer. Each Nacetyl muramic acid is attached to a short (4- to 5-residue) amino acid chain, containing L-alanine, D-glutamic acid, meso-diaminopimelic acid, and D-alanine in the case of *Escherichia coli* (a Gram-negative bacteria) or L-alanine, D-glutamine, L-lysine, and D-alanine with a 5-glycine interbridge between tetrapeptides in the case of *Staphylococcus aureus* (a Gram-positive bacteria). These amino acids, except the L-amino acids, do not occur in proteins and are thought to help to protect against attacks by most peptidases.

Cross-linking between amino acids in different linear amino sugar chains occurs with the help of the enzyme transpeptidase and results in a three-dimensional structure that is strong and rigid. The specific amino acid sequence and molecular structure vary with the bacterial species.

Peptidoglycan serves a structural role in the bacterial cell wall, giving structural strength, as well as counteracting the osmotic pressure of the cytoplasm. Peptidoglycan is also involved in binary fission during bacterial cell reproduction.

The peptidoglycan layer is substantially thicker in Gram-positive bacteria (20 to 80 nanometers) than in Gram-negative bacteria (7 to 8 nanometers), with the attachment of the S-layer. Peptidoglycan forms around 90% of the dry weight of Gram-positive bacteria but only 10% of Gram-negative strains. Thus, presence of high levels of peptidoglycan is the primary determinant of the characterisation of bacteria as gram-positive. In Gram-positive strains, it is important in attachment roles and stereotyping purposes. For both Gram-positive and Gram-negative bacteria, particles of approximately 2 nm can pass through the peptidoglycan.

Structure:

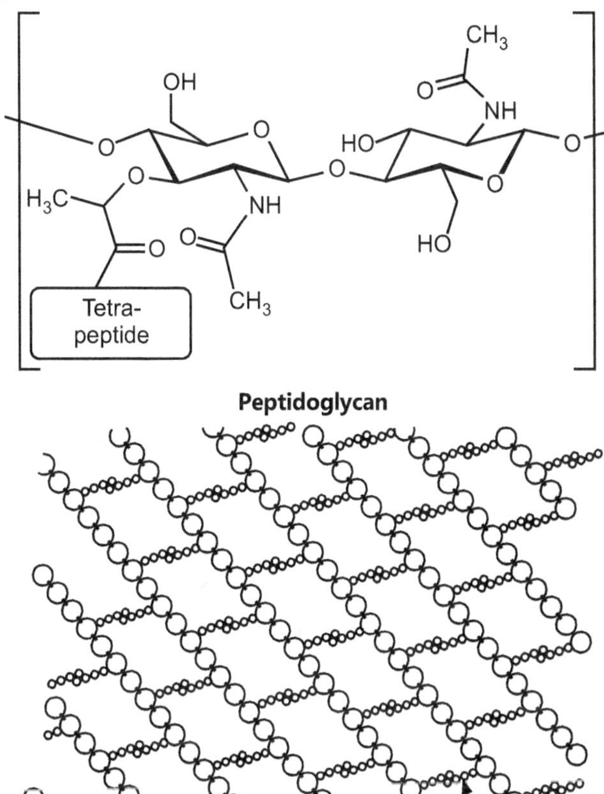

Fig. 2.4: The structure of peptidoglycan

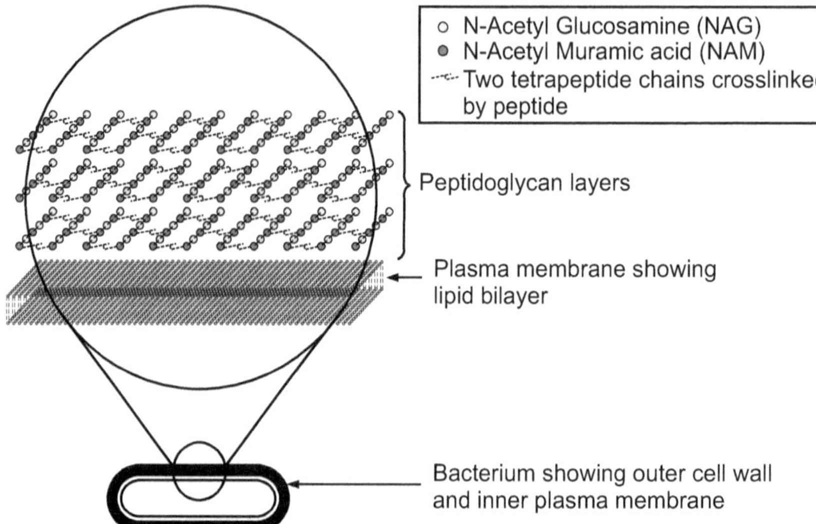

Fig. 2.5: Cell wall in a gram positive bacterium

Chitin is a fibrous substance consisting of polysaccharides, which is the major constituent in the exoskeleton of arthropods and the cell walls of fungi. Chitin ($C_8H_{13}O_5N$), a long-chain polymer of N-acetylglucosamine, is a derivative of glucose. Chitin is a primary component in the exoskeletons of arthropods and crustaceans and is found in the cell walls of fungi. It is a polysaccharide, and it is excreted by the epidermal cells in arthropods.

Chitin is a modified polysaccharide that contains nitrogen; it is synthesized from units of N-acetyl-D-glucosamine (to be precise, 2-(acetylamino)-2-deoxy-D-glucose). These units form covalent β-(1 → 4)-linkages (like the linkages between glucose units forming cellulose). Therefore, chitin may be described as cellulose with one hydroxyl group on each monomer replaced with an acetyl amine group. This allows for increased hydrogen bonding between adjacent polymers, giving the chitin-polymer matrix increased strength.

2.6 LIPIDS

- Lipids are made of the elements Carbon, Hydrogen and Oxygen, but have a much lower proportion of water than other molecules such as carbohydrates.
- Unlike polysaccharides and proteins, lipids are not polymers, they lack a repeating monomeric unit.
- They are made up of two molecules: Glycerol and Fatty acids.
- A glycerol molecule is made up from three carbon atoms with a hydroxyl group attached to it and hydrogen atoms occupying the remaining positions.
- Fatty acids consist of an acid group at one end of the molecule and a hydrocarbon chain, which is usually denoted by the letter 'R'.
- They may be saturated or unsaturated.
- A fatty acid is saturated if every possible bond is made with a Hydrogen atom, such that there exist no $C=C$ bonds.
- Saturated fatty acids on the other hand contain $C=C$ bonds. Monounsaturated fatty acids have one $C=C$ bond, and polyunsaturated have more than one $C=C$ bond.
- Lipids include fats, oils, waxes and related compounds.
- They are widely distributed in nature both in plants and in animals.

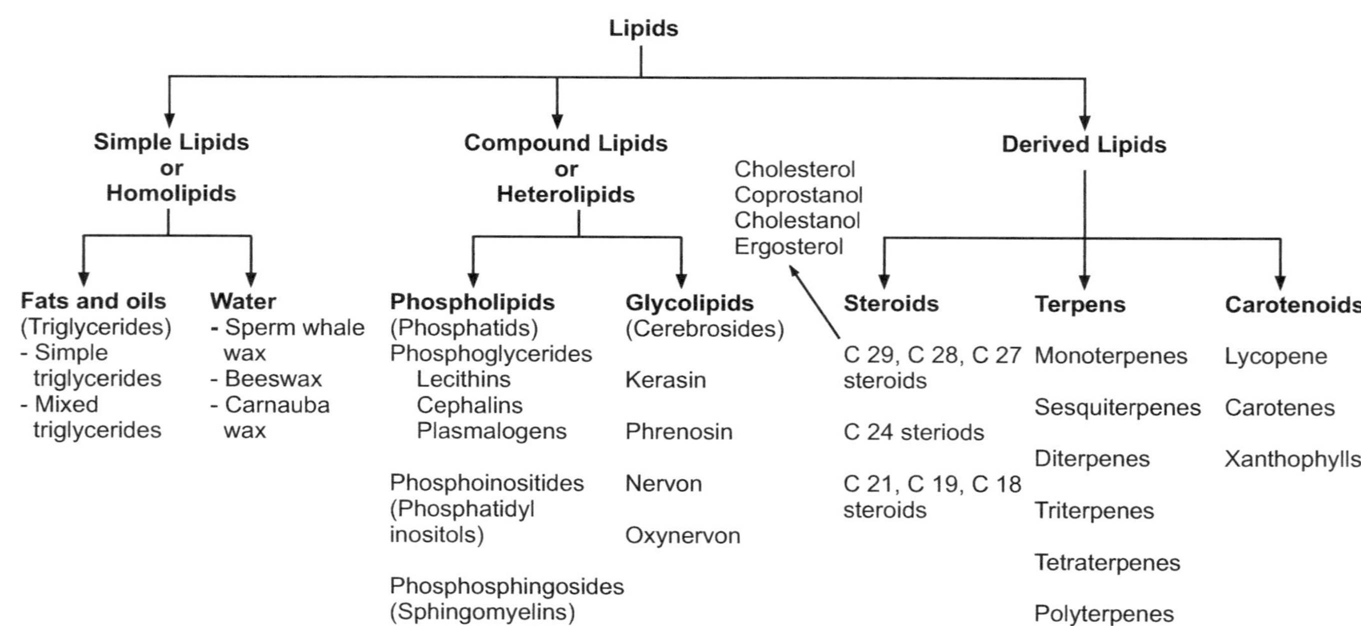

Classification of Lipids

Biological Importance of Lipids:

1. They are more palatable and storable to unlimited amount compared to carbohydrates.
2. They have a high-energy value (25% of body needs) and they provide more energy per gram than carbohydrates and proteins but carbohydrates are the preferable source of energy.
3. Supply the essential fatty acids that cannot be synthesized by the body.
4. Supply the body with fat-soluble vitamins (A, D, E and K).
5. They are important constituents of the nervous system.

Tissue fat is an essential constituent of cell membrane and nervous system. It is mainly phospholipids in nature that are not affected by starvation.

Lipids can be classified according to their hydrolysis products and according to similarities in their molecular structures.

Three major subclasses are recognized:

2.6.1 Simple Lipids

(a) **Triglycerides**

(b) **Fats and oils** which yield fatty acids and glycerol upon hydrolysis.

(c) **Waxes,** which yield fatty acids and long-chain alcohols upon hydrolysis.

(a) Triglycerides:

- They are uncharged due to absence of ionizable groups in it.
- They are the most abundant lipids in nature. They constitute about 98% of the lipids of adipose tissue, 30% of plasma or liver lipids, less than 10% of erythrocyte lipids.
- They are esters of glycerol with various fatty acids. Since the three hydroxyl groups of glycerol are esterified, the neutral fats are also called "Triglycerides".
- Esterification of glycerol with one molecule of fatty acid gives monoglyceride, and that with two molecules gives diglyceride.
  1) **Simple triglycerides:** If the three fatty acids connected to glycerol are of the same type the triglyceride is called simple triglyceride, e.g., tripalmitin.

2) **Mixed triglycerides:** If they are of different types, called mixed triglycerides, e.g., stearo-diolein and palmito-oleo-stearin.

- A triglyceride (TG, triacylglycerol, TAG, or triacylglyceride) is an ester derived from glycerol and three fatty acids. Triglycerides are a blood lipid that enable the bidirectional transfer of adipose fat and blood glucose from the liver. There are many triglycerides: depending on the oil source, some are highly unsaturated and some less.

(b) Fats and Oils:

- Both types of compounds are called triacylglycerols because they are esters composed of three fatty acids joined to glycerol, a trihydroxy alcohol.

- The difference is on the basis of their physical states at room temperature. It is customary to call a lipid and a fat if it is solid at 25°C and oil if it is a liquid at the same temperature.

- These differences in melting points reflect differences in the degree of unsaturation of the constituent fatty acids.

(c) Waxes:

- A wax is an ester of a long-chain alcohol (usually mono-hydroxy) and a fatty acid.

- The acids and alcohols normally found in waxes have chains of the order of 12-34 carbon atoms in length.

2.6.2 Compound Lipids

(a) Phospholipids, which yield fatty acids, glycerol, amino alcohol sphingosine, phosphoric acid and a nitrogen-containing alcohol upon hydrolysis.

They may be glycerophospholipids or sphingophospholipid depending upon the alcohol group present (glycerol or sphingosine).

(b) Glycolipids, which yield fatty acids, sphingosine or glycerol, and a carbohydrate upon hydrolysis.

They may also be glyceroglycolipids or sphingoglycolipid depending upon the alcohol group present (glycerol or sphingosine).

2.6.3 Derived Lipids

Hydrolysis product of simple and compound lipids is called derived lipids. They include fatty acid, glycerol, sphingosine and steroid derivatives.

Steroid derivatives are phenanthrene structure that are quite different from lipids made up of fatty acids.

Steroid is a biologically active organic compound with four rings arranged in a specific molecular configuration. Steroids have two principal biological functions: as important components of cell membranes which alter membrane fluidity; and as signaling molecules. Hundreds of steroids are found in plants, animals and fungi. All steroids are manufactured in cells from the sterols lanosterol (eukaryotes) or cycloartenol (plants). Lanosterol and cycloartenol are derived from the cyclization of the triterpenesqualene.

The steroid core structure is typically composed of seventeen carbon atoms, bonded in four "fused" rings: three six-member cyclohexane rings (rings A, B and C) and one five-member cyclopentane ring (D ring). Steroids vary by the functional groups attached to this four-ring core and by the oxidation state of the rings. Sterols are forms of steroids with a hydroxy group at position three and a skeleton derived from cholestane. Steroids can also be more radically modified, such as by changes to the ring structure, for example, cutting one of the rings. Cutting ring B produces secosteroids one of which is vitamin D_3.

Examples include the lipidcholesterol, the sex hormones estradiol and testosterone and the anti-inflammatory drug dexamethasone.

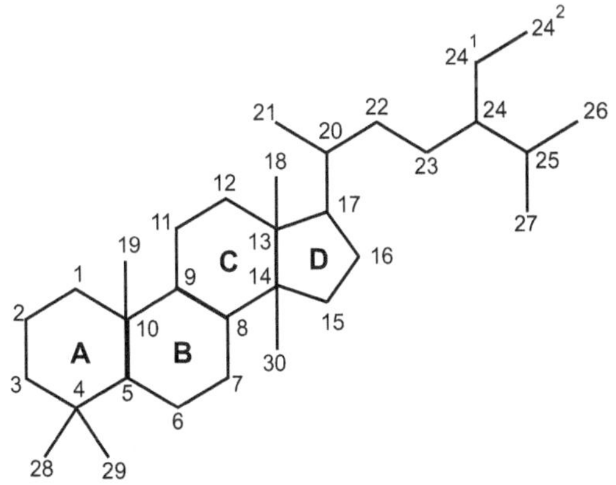

Cholesterol Structure:

Cholesterol is a lipid with a unique structure consisting of four linked hydrocarbon rings forming the bulky steroid structure. There is a hydrocarbon tail linked to one end of the steroid and a hydroxyl group linked to the other end.

Cholesterol also serves as a precursor for the biosynthesis of steroid hormones, bile acid and vitamin D. Cholesterol is the principal sterol synthesized by all animals. In vertebrates, hepatic cells typically produce the greatest amounts.

Structure of cholesterol

2.7 PROTEINS

Proteins are large biomolecules, or macromolecules, consisting of one or more long chains of amino acid residues. Proteins perform a vast array of functions within organisms, including catalysing various metabolic reactions, DNA replication, responding to stimuli, providing structural strength to cells and organisms, and transporting molecules from one part to another. Proteins differ from one another primarily in their sequence of amino acids, which is dictated by the nucleotide sequence of their genes, and which usually results in protein folding into a specific three-dimensional structure that determines its activity.

A linear chain of amino acid residues is called a polypeptide. A protein contains at least one long polypeptide. Short polypeptides, containing less than 20–30 residues, are commonly called peptides, or sometimes oligopeptides. The individual amino acid residues are bonded together by peptide bonds and adjacent amino acid residues. The sequence of amino acid residues in a protein is defined by the sequence of a gene, which is encoded in the genetic code. In general, the genetic code specifies 20 standard amino acids. Proteins can also work together to achieve a particular function.

Once formed, proteins only exist for a certain period and are then degraded and recycled by the cell's machinery. A protein's lifespan is measured in terms of its half-life and covers a wide range. They can exist for minutes or years with an average life span of 1–2 days in mammalian cells.

Proteins are essential parts of organisms and participate in virtually every process within cells. Many proteins are enzymes that catalyse biochemical reactions and are vital to metabolism. Proteins also have structural or mechanical functions, such as actin and myosin in muscle and the proteins in the cytoskeleton, which maintains cell shape. Other proteins are important in cell signaling, immune responses, cell adhesion, and the cell cycle.

Most proteins consist of linear polymers built from series of up to 20 different L-α-amino acids. All proteinogenic amino acids possess common structural features, including an α-carbon to which an amino group, a carboxyl group, and a variable side chain are bonded. Only proline differs from this basic structure as it contains an unusual ring to the N-end amine group, which forces the CO–NH amide moiety into a fixed conformation.

The proteins are classified as structural and functional proteins.

2.7.1 Structural Protein

The largest class is that of structural ones when measured in terms of their total mass. Structural proteins are fibrous proteins. The most familiar of the fibrous proteins are the keratins, which form the protective covering of all land vertebrates: skin, fur, hair, wool, claws, nails, hooves, horns, scales, beaks and feathers. Equally widespread, if less visible, are the actin and myosin proteins of muscle tissue. Another group of fibrous structural proteins are the silks and insect fibres. In addition, there are the collagens of tendons and hides, which form connective ligaments within the body and give extra support to the skin where needed.

2.7.2 Functional Protein

The meaning of a functional protein is that it has the ability to carry out metabolic processes like breakdown (anabolic) and build up (catabolic) tissues for structural integrity, and also help build the cell walls (peptidoglycan) of plants and some bacteria, as well as humans (peripheral and integral proteins) which help to take in and throw out nutrients from the cell.

The tertiary structure of a protein is the functional protein.

Amino acids are molecules used to build proteins. All amino acids have a central carbon atom surrounded by a hydrogen atom, a carboxyl group (COOH), an amino group (NH_2), and an R-group. The R-group or side chain differs between the 20 amino acids.

H   H   O
| | ||
H — N — C — C — OH
|
R

Amino   Carboxyl
group   R   group
Side
chain

Amino acids are organic compounds that contain amine ($-NH_2$) and carboxyl ($-COOH$) functional groups, along with a side chain (R group) specific to each amino acid. The key elements of an amino acid are carbon (C), hydrogen (H), oxygen (O), and nitrogen (N), although other elements are found in the side chains of certain amino acids.

Twenty of the proteinogenic amino acids are encoded directly by triplet codons in the genetic code and are known as "standard" amino acids. The other two ("non-standard" or "non-canonical") are selenocysteine (present in many prokaryotes as well as most eukaryotes, but not coded directly by DNA), and pyrrolysine (found only in some archaea and one bacterium).

Nine proteinogenic amino acids are called "essential" for humans because they cannot be produced from other compounds by the human body and so must be taken in as food. Others may be conditionally essential for certain ages or medical conditions. Essential amino acids may also differ between species.

Peptide Bond:

A peptide bond is a chemical bond formed between two molecules when the carboxyl group of one molecule reacts with the amino group of the other molecule, releasing a molecule of water

(H_2O). This is a dehydration synthesis reaction (condensation reaction), and usually occurs between amino acids.

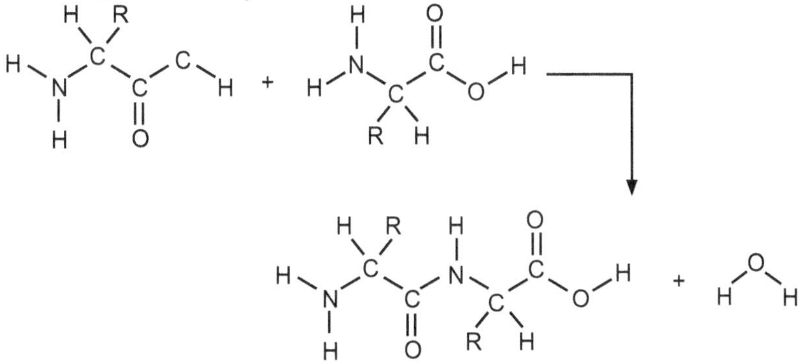

The formation of the peptide bond consumes energy, which, in organisms, is derived from ATP. Peptides and proteins are chains of amino acids held together by peptide bonds.

Amino acids have the following structural properties:

- A carbon (the alpha carbon)
- A hydrogen atom (H)
- A Carboxyl group (–COOH)
- An Amino group (–NH_2)
- A "variable" group or "R" group

Non-polar, polar, and electrically charged are the three properties of side chains used to classify amino acids.

2.7.3 Amino Acid Groups

Amino acids can be classified into four general groups based on the properties of the "R" group in each amino acid. The R group is the side chain of amino acids. Amino acids can be polar, non-polar, positively charged, or negatively charged. Within each class, there are gradations of polarity, size, and shape of the R groups.

Polar amino acids have "R" groups that are hydrophilic, having affinity for water. Non-polar amino acids are the hydrophobic in that they repel water. These interactions play a major role in protein folding and give proteins their 3-D structure. Below is a listing of the 20 amino acids grouped by their "R" group properties. The non-polar amino acids are hydrophobic, while the remaining groups are hydrophilic.

Types of amino acids based on R-groups:

Non-polar Amino Acids:

- **Ala:** Alanine   **Gly:** Glycine   **Ile:** Isoleucine
- **Leu:** Leucine   **Met:** Methionine **Trp:** Tryptophan
- **Phe:** Phenylalanine **Pro:** Proline   **Val:** Valine

Polar Amino Acids:

- **Cys:** Cysteine   **Ser:** Serine   **Thr:** Threonine
- **Tyr:** Tyrosine   **Asn:** Asparagine **Gln:** Glutamine

Polar Basic Amino Acids (Positively Charged):

- **His:** Histidine   **Lys:** Lysine   **Arg:** Arginine

Polar Acidic Amino Acids (Negatively Charged):

- **Asp:** Aspartate   **Glu:** Glutamate

2.7.4 Structural Levels of Proteins

The four levels of protein structure are primary, secondary, tertiary, and quaternary. It is helpful to understand the nature and function of each level of protein structure in order to fully understand how a protein works.

- **Primary structure is the amino acid sequence.**

A protein's primary structure is the unique sequence of amino acids in each polypeptide chain that makes up the protein. Because the final protein structure ultimately depends on this sequence, this is called the primary structure of the polypeptide chain. For example, the pancreatic hormone insulin has two polypeptide chains, A and B.

- **Secondary structure is local interactions between stretches of a polypeptide chain and includes α-helix and β-pleated sheet structures.**

A protein's secondary structure arises from interactions between neighbouring or near-by amino acids as the polypeptide starts to fold into its functional three-dimensional form. Secondary structures arise as H bonds form between local groups of amino acids in a region of the polypeptide chain. The most common forms of secondary structure are the α-helix and β-pleated sheet structures and they play an important structural role in most globular and fibrous proteins.

In the α-helix chain, the hydrogen bond forms between the oxygen atom in the polypeptide backbone carbonyl group in one amino acid and the hydrogen atom in the polypeptide backbone amino group of another amino acid that is four amino acids further along the chain. This holds the stretch of amino acids in a right-handed coil. Every helical turn in an alpha helix has 3.6 amino acid residues. The R groups (the side chains) of the polypeptide protrude out from the α-helix chain and are not involved in the H bonds that maintain the α-helix structure.

In β-pleated sheets, stretches of amino acids are held in an almost fully-extended conformation that "pleats" or zig-zags due to the non-linear nature of single C-C and C-N covalent bonds. β-pleated sheets never occur alone. They have to be held in place by other β-pleated sheets.

- **Tertiary structure is the overall three-dimensional folding driven largely by interactions between R groups.**

The tertiary structure of a polypeptide chain is its overall three-dimensional shape, which occurs once all the secondary structure elements have folded together among each other. Interactions between polar, non-polar, acidic, and basic R group within the polypeptide chain create the complex three-dimensional tertiary structure of a protein. When protein folding takes place in the aqueous environment of the body, the hydrophobic R groups of non-polar amino acids mostly lie in the interior of the protein, while the hydrophilic R groups lie mostly on the outside. Cysteine side chains form disulfide linkages in the presence of oxygen, the only covalent bond formed during protein folding. All of these interactions, weak and strong, determine the final three-dimensional shape of the protein. When a protein loses its three-dimensional shape, it is no longer functional.

- **Quarternary structure is the orientation and arrangement of subunits in a multi-subunit protein.**

The quaternary structure of a protein is how its subunits are oriented and arranged with respect to one another. As a result, quaternary structure only applies to multi-subunit proteins; that is, proteins made from more than one polypeptide chain. Proteins made from a single polypeptide will not have a quaternary structure.

In proteins with more than one subunit, weak interactions between the subunits help to stabilize the overall structure. Enzymes often play key roles in bonding subunits to form the final functioning protein.

For example, insulin is a ball-shaped, globular protein that contains both hydrogen bonds and disulfide bonds that hold its two polypeptide chains together. Silk is a fibrous protein that results from hydrogen bonding between different β-pleated chains.

2.7.5 Haemoglobin

Haemoglobin (Hb) is the iron-containing oxygen-transport metalloprotein in the red blood cells of all vertebrates as well as the tissues of some invertebrates. Haemoglobin in the blood carries oxygen from the respiratory organs (lungs or gills) to the rest of the body (i.e. the tissues) where it releases the oxygen to burn nutrients to provide energy to power the functions of the organism, and collects the resultant carbon dioxide to bring it back to the respiratory organs to be dispensed from the organism.

In mammals, the protein makes up about 97% of the red blood cell's dry content (by weight), and around 35% of the total content (including water). Haemoglobin has an oxygen binding capacity of 1.34 ml O_2 per gram of haemoglobin, which increases the total blood oxygen capacity seventy-fold compared to dissolved oxygen in blood. The mammalian haemoglobin molecule can bind up to four oxygen molecules.

Haemoglobin is involved in the transport of other gases: it carries some of the body's respiratory carbon dioxide (about 10% of the total) as carbaminohemoglobin, in which CO_2 is bound to the globin protein. The molecule also carries the important regulatory molecule nitric oxide bound to a globin protein thiol group, releasing it at the same time as oxygen.

Haemoglobin is also found outside red blood cells. Other cells that contain haemoglobin include the A9 dopaminergic neurons in the substantia nigra, macrophages, alveolar cells, and mesangial cells in the kidney. In these tissues, haemoglobin has a non-oxygen-carrying function as an antioxidant and a regulator of iron metabolism.

Haemoglobin and haemoglobin-like molecules are also found in many invertebrates, fungi, and plants. In these organisms, haemoglobin may carry oxygen, or they may act to transport and regulate other things such as carbon dioxide, nitric oxide, hydrogen sulfide. A variant of the molecule, called leghemoglobin, is used to scavenge oxygen away from anaerobic systems, such as the nitrogen-fixing nodules of leguminous plants, before the oxygen can poison the system.

Structure:

At the core of the molecule is a heterocyclic ring, known as a porphyrin which holds an iron atom; this iron atom is the site of oxygen binding. An iron containing porphyrin is termed a haeme. The name haemoglobin is the conjugate of haeme and globin, a globin being a generic term for a globular protein. There are a number of heme containing proteins haemoglobin is the best known.

In adult humans, haemoglobin is a tetramer (contains 4 subunit proteins), consisting of two alpha and two beta subunits which are non-covalently bound. The subunits are structurally similar and about the same size. Each subunit has a molecular weight of about 16,000 Daltons, for a total molecular weight in the tetramer of about 64,000 Daltons. Each subunit of haemoglobin contains a single haeme, so that the overall binding capacity of adult human haemoglobin for oxygen is four oxygen molecules:

Stepwise Reaction:

- $Hb + O_2 \longleftrightarrow HbO_2$
- $HbO_2 + O_2 \longleftrightarrow Hb(O_2)_2$
- $Hb(O_2)_2 + O_2 \longleftrightarrow Hb(O_2)_3$
- $Hb(O_2)_3 + O_2 \longleftrightarrow Hb(O_2)_4$

Summary Reaction:

- $Hb + 4O_2 \longrightarrow Hb(O_2)$

Flagellin:

Flagellin is a globular protein that arranges itself in a hollow cylinder to form the filament in a bacterial flagellum. It has a mass of about 30,000 to 60,000 daltons. Flagellin is the principal component of bacterial flagellum, and is present in large amounts on nearly all flagellated bacteria.

The structure of flagellin is responsible for the helical shape of the flagellar filament, which is important for its proper function. It is transported through the center of the filament to the tip where it polymerases spontaneously into a part of the filament. The filament is made up of eleven smaller "protofilaments", nine of which contains flagellin in the L-type shape and the other two in the R-type shape.

The helical N- and C-termini of flagellin form the inner core of the flagellin protein, and is responsible for flagellin's ability to polymerize into a filament. The middle residues make up the outer surface of the flagellar filament. While the termini of the protein are quite similar among all bacterial flagellins, the middle portion is wildly variable and can be absent in some species. The flagellin domains are numbered from the helical core (D_0/D_1) to the outside (D_2 ...); when viewed from the amino-acid sequence, D_0/D_1 appears on the two termini. Flagellin-like structural proteins are found in other portions of the flagellum, such as the hook, the rod at the base, and the cap at the top.

The middle part of *E. coli* (and related) flagellin, D_3, displays a beta-folium fold and appears to maintain flagellar stability.

Cytoskeletal proteins are proteins that make up the cytoskeleton, flagella or cilia of cells. Through a series of intercellular proteins, the cytoskeleton gives a cell its shape, offers support, and facilitates movement through three main components: microfilaments, intermediate filaments, and microtubules. The cytoskeleton has a variety functions including, giving shape to cells lacking a cell wall, allowing for cell movement, enabling movement of organelles within the cell, endocytosis, and cell division. Generally, cytoskeletal proteins are polymers, and include tubulin (the protein component of microtubules), actin (the component of microfilaments) and lamin (the component of intermediate filaments).

Tubulin:

Tubulinrefers either to the tubulin protein superfamily of globular proteins, or one of the member proteins of that superfamily. α- and β-tubulins polymerize into microtubules, a major component of the eukaryotic cytoskeleton. Microtubules function in many essential cellular processes, including mitosis.

In eukaryotes, there are six members of the tubulin super family, although not all are present in all species. Both α and β tubulins have a mass of around 50 kDa and are thus in a similar range compared to actin (with a mass of ~42 kDa). In contrast, tubulin polymers (microtubules) tend to be much bigger than actin filaments due to their cylindrical nature.

Actin:

Actin is a family of globular multi-functional proteins that form microfilaments. It is found in essentially all eukaryotic cells where it may be present at a concentration of over 100 μM; its mass is roughly 42-kDa, with a diameter of 4 to 7 nm.

Actin is a key component of the cytoskeleton in all eukaryotic organisms. It is critical for cell movement, determination of cell shape and cell division, and it plays important roles in many other processes like cell structural support, axonal growth, cell migration, organelle transport and phagocytosis.

Lamin:

Lamins are fibrous proteins providing structural function and transcriptional regulation in the cell nucleus. Nuclear lamins interact with inner nuclear membrane proteins to form the nuclear lamina on the interior of the nuclear envelope.

The nuclear lamina is a dense (~30 to 100 nm thick) fibrillar network inside the nucleus of most cells. It is composed of intermediate filaments and membrane associated proteins. Besides providing mechanical support, the nuclear lamina regulates important cellular events such as DNA replication and cell division.

2.8 NUCLEIC ACIDS

A nucleic acid is a chain of nucleotides which stores genetic information in biological systems. The functions of nucleic acids are the storage and expression of genetic information. Deoxyribonucleic acid (DNA) encodes the information the cell needs to make proteins. A related type of nucleic acid, called ribonucleic acid (RNA), comes in different molecular forms that participate in protein synthesis. This information is stored in multiple sets of three nucleotides, known as codons.

Nucleic acids are the biopolymers, or small biomolecules, essential to all known forms of life. The term nucleic acid is the overall name for DNA and RNA. They are composed of nucleotides, which are the monomers made of three components: a 5-carbon sugar, a phosphate group and a nitrogenous base. If the sugar is a compound ribose, the polymer is RNA (ribonucleic acid); if the sugar is derived from ribose as deoxyribose, the polymer is DNA (deoxyribonucleic acid).

Nucleic acids are the most important of all biomolecules. These are found in abundance in all living things, where they function to create and encode, and then store information of every living cell of every life-form organism on Earth. In turn, they function to transmit and express that information inside and outside the cell nucleus, to the interior operations of the cell and ultimately to the next generation of each living organism. The encoded information is contained and conveyed via the nucleic acid sequence, which provides the 'ladder-step' ordering of nucleotides within the molecules of RNA and DNA.

Strings of nucleotides are bonded to form helical backbones typically, one for RNA, two for DNA and assembled into chains of base-pairs selected from the five primary, nucleobases, which are: adenine, cytosine, guanine, thymine, and uracil. Thymine occurs only in DNA and uracil only in RNA. Using amino acids and the process known as protein synthesis, the specific sequencing in DNA of these nucleobase-pairs enables storing and transmitting coded instructions as genes. In RNA, base-pair sequencing provides for manufacturing new proteins that determine the frames and parts and most chemical processes of all life forms.

Nucleic acids are generally very large molecules. DNA molecules are probably the largest individual molecules known. The nucleic acid molecules range in size from 21 nucleotides to large chromosomes. In most cases, naturally occurring DNA molecules are double-stranded and RNA molecules are single-stranded. There are numerous exceptions, however, some viruses have genomes made of double-stranded RNA and other viruses have single-stranded DNA genomes and, in some circumstances, nucleic acid structures with three or four strands can form.

Nucleic acids are linear polymers of nucleotides. Each nucleotide consists of three components: a purine or pyrimidine (sometimes termed nitrogenous base or simply base), a pentose sugar, and a phosphate group. The structure consisting of a nucleobase plus sugar is termed a nucleoside. Nucleic acid types differ in the structure of the sugar in their nucleotides–DNA contains 2'-deoxyribose while RNA contains ribose (where the only difference is the presence of a hydroxyl group). Also, the nucleobases found in the two nucleic acid types are different: adenine, cytosine, and guanine are found in both RNA and DNA, while thymine occurs in DNA and uracil occurs in RNA.

The sugars and phosphates in nucleic acids are connected to each other in an alternating chain (sugar-phosphate backbone) through phosphodiester linkages. In conventional nomenclature, the carbons to which the phosphate groups attach are the 3'-end and the 5'-end carbons of the sugar. This gives nucleic acids directionality, and the ends of nucleic acid molecules are referred to as 5'-end and 3'-end. The nucleobases are joined to the sugars via an N-glycosidic linkage involving nucleobase ring nitrogen (N-1 for pyrimidines and N-9 for purines) and the 1' carbon of the pentose sugar ring.

Non-standard nucleosides are also found in both RNA and DNA and usually arise from modification of the standard nucleosides within the DNA molecule or the primary (initial) RNA transcript. Transfer RNA (tRNA) molecules contain a particularly large number of modified nucleosides.

2.8.1 DNA

Deoxyribonucleic acid (DNA) is a molecule composed of two chains that coil around each other to form a double helix carrying genetic instructions for the development, functioning, growth and reproduction of all known organisms and many viruses.

The two DNA strands are also known as polynucleotides as they are composed of simpler monomeric units called nucleotides. Each nucleotide is composed of one of four nitrogen-containing nucleobases (cytosine [C], guanine [G], adenine [A] or thymine [T]), a sugar called deoxyribose, and a phosphate group. The nucleotides are joined to one another in a chain by covalent bonds between the sugar of one nucleotide and the phosphate of the next, resulting in an alternating sugar-phosphate backbone. The nitrogenous bases of the two separate polynucleotide strands are bound together, according to base pairing rules (A with T and C with G), with hydrogen bonds to make double-stranded DNA. The complementary nitrogenous bases are divided into two groups, pyrimidines and purines. In DNA, the pyrimidines are thymine and cytosine; the purines are adenine and guanine.

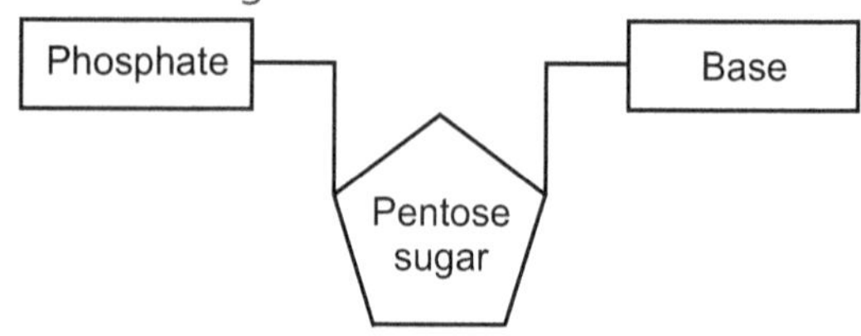

There are four organic bases involved in the formation of DNA molecules:

- adenine and guanine (both purines containing two rings in their structures)
- thymine and cytosine (both pyrimidines containing only one ring in their structures)

Adenine (A)

A and G are double-ring purine bases.

Guanine (G)

Thymine (T)

T and C are single-ring pyrimidine bases.

Cytosine (C)

Both strands of double-stranded DNA store the same biological information. This information is replicated as and when the two strands separate. A large part of DNA (more than 98% for humans) is non-coding, meaning that these sections do not serve as patterns for protein sequences. The two strands of DNA run in opposite directions to each other and are thus antiparallel. Attached to each sugar is one of four types of nucleobases. It is the sequence of these four nucleobases along the backbone that encodes genetic information.

DNA is a long polymer made from repeating units called nucleotides. In all species, it is composed of two helical chains, bound to each other by hydrogen bonds. Both chains are coiled around the same axis, and have the same pitch of 34 angstroms (Å) (3.4 nanometres). The pair of chains has a radius of 10 angstroms (1.0 nanometre).

DNA does not usually exist as a single strand, but instead as a pair of strands that are held tightly together. These two long strands coil around each other, in the shape of a double helix. The nucleotide contains both a segment of the backbone of the molecule (which holds the chain together) and a nucleobase (which interacts with the other DNA strand in the helix). A nucleobase linked to a sugar is called a nucleoside, and a base linked to a sugar and to one or more phosphate groups is called a nucleotide. A biopolymer comprising multiple linked nucleotides (as in DNA) is called a polynucleotide.

The backbone of the DNA strand is made from alternating phosphate and sugar residues. The sugar in DNA is 2-deoxyribose, which is a pentose (five-carbon) sugar. The sugars are joined together by phosphate groups that form phosphodiester bonds between the third and fifth carbon atoms of adjacent sugar rings. These are known as the 3'-end (three prime end), and 5'-end (five prime end) carbons, the prime symbol being used to distinguish these carbon atoms from those of the base to which the deoxyribose forms a glycosidic bond.

Any DNA strand therefore normally has one end at which there is a phosphoryl attached to the 5' carbon of a ribose (the 5' phosphoryl) and another end at which there is a free hydroxyl attached to the 3' carbon of a ribose (the 3' hydroxyl). The orientation of the 3' and 5' carbons along the sugar-phosphate backbone confers polarity to each DNA strand. In a nucleic acid double helix, the direction of the nucleotides in one strand is opposite to their direction in the other strand: the strands are antiparallel. The asymmetric ends of DNA strands are said to have a polarity of five prime end (5').

The DNA double helix is stabilized primarily by two forces: hydrogen bonds between nucleotides and base-stacking interactions among aromatic nucleobases. The four bases found in DNA are adenine (A), cytosine (C), guanine (G) and thymine (T). These four bases are attached to the sugar-phosphate to form the complete nucleotide, as shown for adenosine monophosphate. Adenine pairs with thymine and guanine pairs with cytosine, forming A-T and G-C base pairs.

The nucleobases are classified into two types: the purines, A and G, which are fused five- and six-membered heterocyclic compounds, and the pyrimidines, the six-membered rings C and T.

There are four areas in which the structural forms of DNA can differ.

1. Handedness - right or left
2. Length of the helix turn
3. Number of base pairs per turn
4. Difference in size between the major and minor grooves

The tertiary arrangement of DNA's double helix in space includes B-DNA, A-DNA and Z-DNA.

(a) B-DNA is the most commons form of DNA *in-vivo* and is narrower, elongated helix than A-DNA. Its wide major groove makes it more accessible to proteins. On the other hand, it has a narrow minor groove. B-DNAs favoured conformations occur at high water concentrations and the hydration of the minor groove appears to favour B-DNA. B-DNA base pairs are nearly perpendicular to helix axis.

(b) A-DNA is shorter and wider than helix B. Most RNA and RNA-DNA duplex are in this form. A-DNA has a deep, narrow major groove which does not make it easily accessible to proteins. On the other hand, its wide, shallow minor groove makes it accessible to proteins but with lower information content than the major groove. Its favoured conformation is at low water concentrations. A-DNAs base pairs tilt to helix axis and are displaced from axis.

(c) Z-DNA is a relatively rare left-handed double-helix. Given the proper sequence and superhelical tension, it can be formed *in-vivo* but its function is unclear. It has a narrower, more elongated helix than A or B. Z-DNA's major groove is not really groove and it has a narrow minor groove. The most favoured conformation occurs when there are high salt concentrations. There are some base substitutions but requires an alternating purine-pyrimidine sequence.

2.8.2 RNA

RNA consists of ribose nucleotides (nitrogenous bases appended to a ribose sugar) attached by phosphodiester bonds, forming strands of varying lengths. The nitrogenous bases in RNA are adenine, guanine, cytosine, and uracil, which replaces thymine in DNA.

The ribose sugar of RNA is a cyclical structure consisting of five carbons and one oxygen. The presence of a chemically reactive hydroxyl (–OH) group attached to the second carbon group in the ribose sugar molecule makes RNA prone to hydrolysis. This chemical lability of RNA, compared with DNA, which does not have a reactive –OH group in the same position on the sugar moiety (deoxyribose)

Of the many types of RNA, the three most well-known and most commonly studied are messenger RNA (mRNA), transfer RNA (tRNA), and ribosomal RNA (rRNA), which are present in all organisms. These and other types of RNAs primarily carry out biochemical reactions, similar to enzymes. Some, however, also have complex regulatory functions in cells. Owing to their involvement in many regulatory processes, to their abundance, and to their diverse functions, RNAs play important roles in both normal cellular processes and diseases.

There are three types of RNA, each encoded by its own type of gene:

o   mRNA - Messenger RNA: Encodes amino acid sequence of a polypeptide.

o   tRNA - Transfer RNA: Brings amino acids to ribosomes during translation.

o   rRNA - Ribosomal RNA: With ribosomal proteins, makes up the ribosomes, the organelles that translate the mRNA.

In protein synthesis, mRNA carries genetic codes from the DNA in the nucleus to ribosomes, the sites of protein translation in the cytoplasm. Ribosomes are composed of rRNA and protein. The ribosome protein subunits are encoded by rRNA and are synthesized in the nucleolus. Once fully assembled, they move to the cytoplasm, where, as key regulators of translation, they "read" the code carried by mRNA. A sequence of three nitrogenous bases in mRNA specifies incorporation of a specific amino acid in the sequence that makes up the protein. Molecules of tRNA (sometimes also called soluble, or activator, RNA), which contain fewer than 100 nucleotides, bring the specified amino acids to the ribosomes, where they are linked to form proteins.

(a) mRNA:

Messenger ribonucleic acids (mRNAs) transfer the information from DNA to the cell machinery that makes proteins. Molecules of mRNA are composed of relatively short, single strands of molecules made up of adenine, cytosine, guanine and uracil bases held together by a sugar phosphate backbone. When RNA polymerase finishes reading a section of the DNA, the pre-mRNA copy is processed to form mature mRNA and then transferred out of the cell nucleus.

The mRNA encodes amino acid sequence of a polypeptide.

The primary function of mRNA is to act as an intermediary between the genetic information in DNA and the amino acid sequence of proteins. mRNA contains codons that are complementary to the sequence of nucleotides on the template DNA and direct the formation of amino acids through the action of ribosomes and tRNA. mRNA also contains multiple regulatory regions that can determine the timing and rate of translation. In addition, it ensures that translation proceeds in an orderly fashion because it contains sites for the docking of ribosomes, tRNA as well as various helper proteins.

(b) tRNA:

The tRNA molecule has a distinctive folded structure with three hairpin loops that form the shape of a three-leafed clover. One of these hairpin loops contains a sequence called the anticodon, which can recognize and decode an mRNA codon. Each tRNA has its corresponding amino acid attached to its end.

The tRNA is responsible for bringing amino acids to the ribosome.

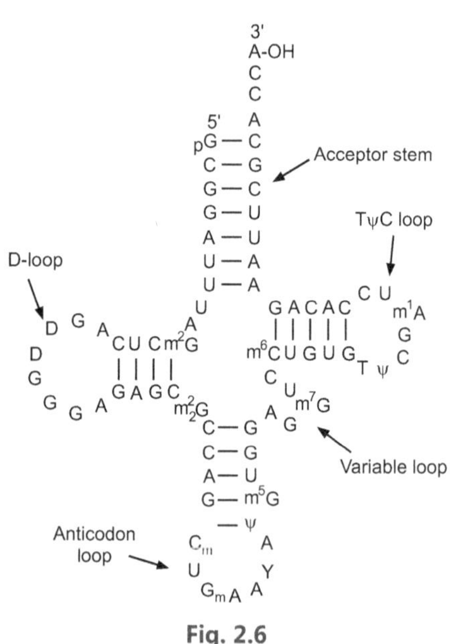

Fig. 2.6

(c) rRNA:

Ribosomal ribonucleic acid (rRNA) is the RNA component of ribosomes, the molecular machines that catalyze protein synthesis. Ribosomal RNA constitute over sixty percent of the ribosome by weight and are crucial for all its functions, from binding to mRNA and recruiting tRNA to catalyzing the formation of a peptide bond between two amino acids. Even the structure of a ribosome is determined by the three-dimensional shape of its rRNA core. Proteins present in the ribosome serve to stabilize this structure through interactions with the core.

The rRNA with ribosomal proteins, makes up the ribosomes, the organelles that translate the mRNA. The primary function of rRNA is in protein synthesis, in binding to messenger RNA and transfer RNA to ensure that the codon sequence of the mRNA is translated accurately into amino acid sequence in proteins. To achieve this, rRNA has a distinctive three-dimensional shape involving internal loops and helices that creates specific sites within the ribosome: the A, P and E sites. The P site is for binding a growing polypeptide, the A site anchors an incoming tRNA charged with an amino acid. After peptide bond formation, the tRNA binds briefly to the E site before leaving the ribosome. In addition rRNA also has sites for binding to some ribosomal proteins and careful analysis has demarcated the exact residues in both the RNA and protein.

Translation of the mRNA sequence requires the involvement of rRNA at every step – initiation, elongation and termination.

2.9 CLASSIFICATION OF BACTERIA

David Hendricks Bergey

David Hendricks Bergey was an American bacteriologist, born in December 27, 1860 in Skippack, Pennsylvania, died September 5, 1937 in Philadelphia, Pennsylvania.

He studied at University of Pennsylvania, where he obtained his Bachelor of Science and Doctor of Medicine degrees in 1884. He practiced medicine until 1893. He then joined the university's hygiene laboratory, where he taught hygiene and bacteriology. He led the laboratory from 1929 until his retirement in 1932.

He was chairman of the Editorial Board for the first edition of Bergey's Manual of Determinative Bacteriology, published in 1923.

The advantages of classifying organisms are as follows:

(i)   Classification facilitates the identification of organisms.

(ii) Helps to establish the relationship among various groups of organisms.

(iii) Helps to study the phylogeny and evolutionary history of organisms.

With the development of bacteriology and a continuous need to identify and classify bacteria various attempts were continuously being made. An important milestone was put forth as the Bergey's manual.

Bergey's manual, was first published in 1923. At present it is in its 9th edition under the title Bergey's Manual of Systematic Bacteriology, It provides a major taxonomic treatment of bacteria. This manual has served the community of microbiologists since more than 90 years and is a collection of information on all recognized species of bacteria.

Many schemes for identification of bacteria were devised prior to 1923 but all were usually fragmentary. There was need for a single scheme which could cover all the described bacteria. David Hendricks Bergey, a professor of bacteriology at the University of Pennsylvania (USA), proceeded in this direction and began preparing a complete review of the enormous literature of bacterial taxonomy.

To aid the publication of this work, the Society of American Bacteriologists (now called the American Society of Microbiologists) appointed an Editorial Board headed by Bergey. This resulted in the publication of the first edition of Bergey's Manual of Determinative Bacteriology in 1923. The second edition of the manual was published in 1925 and the third edition in 1930.

In 1934 The Bergey's Manual Trust was created. Since then, this trust continues to prepare and publish successive editions of the manual and promotes research in the field of bacterial taxonomy.

The first eight editions of this manual appeared under the title 'Bergey's Manual of Determinative Bacteriology'. The 9th edition was retitled and was published as the 1st edition under the title 'Bergey's Manual of Systematic Bacteriology', which consisted of four volumes published in 1984, 1986, 1989, and 1991, respectively. It now aimed at a systematic classification of bacteria.

Volume 1 included information on all types of Gram-negative bacteria that were considered to have "medical and industrial importance." Volume 2 included information on all types of Gram-positive bacteria. Volume 3 deals with all of the remaining, slightly different Gram-negative bacteria, along with the Archaea. Volume 4 has information on filamentous actinomycetes and other, similar bacteria.

The current volumes differ drastically from previous volumes in that many higher taxa are not defined in terms of phenotype, but solely on 16S phylogeny, as is the case of the classes within Proteobacteria.

The current grouping is:

- Volume 1 (2001): The Archaea and the deeply branching and phototrophic Bacteria
- Volume 2 (2005): The Proteobacteria—divided into three books:
  - o 2A: Introductory essays
  - o 2B: The Gamma proteobacteria
  - o 2C: Other classes of Proteobacteria
- Volume 3 (2009): The Firmicutes
- Volume 4 (2011):
  The Bacteroidetes, Spirochaetes, Tenericutes (Mollicutes), Acidobacteria, Fibrobacteres, Fusobacteria, Dictyoglomi, Gemmatimonadetes, Lentisphaerae, Verrucomicrobia, Chlamydiae, and Planctomycetes
- Volume 5 (in two parts) (2012): The Actinobacteria

Bergey's manual of systematics of archaea and bacteria (2015), an online book, replaces the five-volume set.

2.10 CLASSIFICATION OF VIRUSES (ICTV NOMENCLATURE)

The **International Committee on Taxonomy of Viruses (ICTV)** authorizes and organizes the taxonomic classification and the nomenclatures for viruses. The ICTV have developed a universal taxonomic-scheme for viruses. This provides a means to describe, name, and classify every virus that affects living organisms.

The ICTV's essential principles of virus nomenclature are:

- Stability
- To avoid or reject the use of names which might cause error or confusion
- To avoid the unnecessary creation of names

Species: A species name shall consist of as few words as practicable but must not consist only of a host name and the word virus. A species name must provide an appropriately unambiguous identification of the species. Numbers, letters, or combinations thereof may be used as species epithets where such numbers and letters are already widely used.

Genera: A virus genus is a group of related species that share some significant properties and often only differ in host range and virulence. A genus name must be a single word ending in the suffix - virus.

Subfamilies: A subfamily is a group of genera sharing certain common characters. A subfamily name must be a single word ending in the suffix - *virinae*.

Families: A family is a group of genera, whether or not these are organized into subfamilies, sharing certain common characters. A family name must be a single word ending in the *suffix - viridae*.

Orders: An order is a group of families sharing certain common characters. An order name must be a single word ending in the suffix - virales.

Rules for sub-viral agents:

Rules concerned with the classification of viruses shall also apply to the classification of viroids. The formal endings for taxa of viroids are the word viroid for species, the suffix - *viroid* for genera, the suffix - *viroinae* for sub-families, should this taxon be needed, and - *viroidae* for families.

Retrotransposons are considered to be viruses in classification and nomenclature. Satellites and prions are not classified as viruses but are assigned an arbitrary classification as seems useful to workers in the particular fields.

Rules for Orthography:

1. In formal taxonomic usage the accepted names of virus orders, families, subfamilies, and genera are printed in italics and the first letters of the names are capitalized.

2. Species names are printed in italics and have the first letter of the first word capitalized. Other words are not capitalized unless they are proper nouns, or parts of proper nouns.

3. In formal usage, the name of the taxon shall precede the term for the taxonomic unit.

Viral classification starts at the level of realm and continues as follows, with the taxon suffixes given in italics

Realm *(-viria)*

Subrealm *(-vira)*

Kingdom *(-viriae)*

Subkingdom *(-virites)*

Phylum *(-viricota)*

Subphylum *(-viricotina)*

Class *(-viricetes)*

Subclass *(-viricetidae)*

Order *(-virales)*

Suborder *(-virineae)*

Family *(-viridae)*

Subfamily *(-virinae)*

Genus *(-virus)*

Subgenus *(-virus)*

Species

Species names often take the form of [*Disease*] virus, particularly for higher plants and animals. As of November 2018, only phylum, subphylum, class, order, suborder, family, subfamily, genus, and species are used.

The establishment of an order is based on the inference that the virus families it contains have most likely evolved from a common ancestor. The majority of virus families remain unplaced.

As of 2018, one realm, four incertaesedis orders, 46 incertaesedis families, and three incertaesedis genera are accepted.

Realms: Riboviria

Incertaesedis orders: *Caudovirales, Herpesvirales, Ligamenvirales, Ortervirales.*

Incertaesedis families: *Adenoviridae, Alphasatellitidae, Ampullaviridae, Anelloviridae, Ascoviridae, Asfarviridae, Bacilladnaviridae, Baculoviridae, Bicaudaviridae, Bidnaviridae, Circoviridae, Clavaviridae, Corticoviridae, Fuselloviridae, Geminiviridae, Genomoviridae, Globuloviridae, Guttaviridae, Hepadnaviridae, Hytrosaviridae, Inoviridae, Iridoviridae, Lavidaviridae, Marseilleviridae, Microviridae, Mimiviridae, Nanoviridae, Nimaviridae, Nudiviridae, Ovaliviridae, Papillomaviridae, Parvoviridae, Phycodnaviridae, Plasmaviridae, Pleolipoviridae, Polydnaviridae, Polyomaviridae, Portogloboviridae, Poxviridae, Smacoviridae, Sphaerolipoviridae,*

Spiraviridae,   Tectiviridae,   Tolecusatellitidae,   Tristromaviridae, Turriviridae.

Incertaesedis genera: *Dinodnavirus, Rhizidiovirus, Salterprovirus.*

Higher virus taxa span viruses with varying host ranges. The *Ortervirales* Groups VI and VII), containing also retroviruses (infecting animals including humans e.g. HIV), retrotransposons (infecting invertebrate animals, plants and eukaryotic microorganisms) and caulimoviruses (infecting plants), are recent additions to the classification system orders. Other variations occur between the orders: Nidovirales, for example, are isolated for their differentiation in expressing structural and non-structural proteins separately.

Think Over It

1.   What makes lipids different from other biomolecules?
2.   Cellulose is composed of which basic units?
3.   What is a ribose nucleotide made up of ?
4.   What is ATP?
5.   Is DNA directly involved in protein synthesis?
6.   Which bonds maintain the tertiary structure of proteins?
7.   Why was the 9th edition called as the systematic bacteriology?
8.   Why the basis for bacterial and viral classification needs to be different?

SUMMARY

* An atom is a fundamental piece of matter that uniquely defines a chemical element.
* Amino acids, lipids, carbohydrates, proteins, polysaccharides, and nucleic acids as well as small molecules such as primary metabolites, secondary metabolites, and natural products are included in biomolecules.
* Various types of bonds like the covalent bond, non-covalent bond, ester bond, Phospho-diester bond, Peptide bond and Glycosidic bond are present in the biomolecules.
* Carbohydrates are compounds made from three elements: carbon, hydrogen and oxygen.

- A glycosidic bond or glycosidic linkage is a type of covalent bond that joins a carbohydrate (sugar) molecule to another group, which may or may not be another carbohydrate.
- Monosaccharides (e.g. glucose) and disaccharides (e.g. sucrose) are relatively small molecules. Other carbohydrate molecules are very large (polysaccharides such as starch and Monosaccharides containing the aldehyde group are classified as aldoses, and those with a ketone group are classified as ketoses.
- Lipids are made of the elements Carbon, Hydrogen and Oxygen, but have a much lower proportion of water. Lipids are not polymers, they lack a repeating monomeric unit.
- They are made from two molecules: Glycerol and Fatty acids.
- Lipids can be classified as Simple lipids (Triglycerides, Fats and oils, waxes), Compound lipids (Phospholipid, Glycolipids) and Derived lipids (Steroids, Cholesterol).
- Proteins are large biomolecules, or macromolecules, consisting of one or more long chains of amino acid residues. Proteins perform a vast array of functions within organisms, including catalysing various metabolic reactions.
- The proteins are classified as structural and functional proteins.
- Amino acids are molecules used to build proteins. All amino acids have a central carbon atom surrounded by a hydrogen atom, a carboxyl group (COOH), an amino group (NH_2), and an R-group. Amino acids are classified based on R group.
- A peptide bond is a chemical bond formed between two molecules when the carboxyl group of one molecule reacts with the amino group of the other molecule, releasing a molecule of water (H_2O). This is a condensation reaction, and usually occurs between amino acids.
- There are four structural levels of proteins: primary, secondary, tertiary and quaternary.
- A nucleic acid is a chain of nucleotides which stores genetic information in biological systems. Deoxyribonucleic acid (DNA) encodes the information the cell needs to make proteins. Ribonucleic acid (RNA) comes in different molecular forms that participate in protein synthesis.

- Deoxyribonucleic acid (DNA) is a molecule composed of two chains that coil around each other to form a double helix carrying genetic instructions for the development, functioning, growth and reproduction of all known organisms and many viruses.
- Ribonucleic acid (RNA) primarily carry out biochemical reactions, similar to enzymes and consists of ribose nucleotides (nitrogenous bases appended to a ribose sugar) attached by phosphodiester bonds, forming strands of varying lengths.
- There are three types of RNA: m-RNA, t-RNA, r-RNA.
- Bergey's manual, was first published in 1923. The 9th edition was retitled and was published as the 1st edition under the title 'Bergey's Manual of Systematic Bacteriology'
- The current grouping is in Five Volumes.
- The **International Committee on Taxonomy of Viruses (ICTV)** authorizes and organizes the taxonomic classification and the nomenclatures for viruses.
- Viral classification starts at the level of realm and continues, with the taxon suffixes given in italics.

Exercise

(A) Questions for One Mark:

1. What is an atom?
2. What are the different types of biomolecules?
3. Which types of bonds are present in biomolecules?
4. What is covalent bond?
5. What is a non-covalent bond?
6. What is a ester bond?
7. What is a phospho-diester bond?
8. What is a peptide bond?
9. What is a glycosidic bond?
10. Which elements make a carbohydrate?
11. What are monosaccharides?
12. What are disaccharides?
13. What are oligosaccharides?
14. What are polysaccharides?
15. What are lipids made up of?
16. Give examples of simple lipids/compound lipids/derived lipids.
17. What are steroids?

18. What are proteins?
19. Give two examples of structural proteins.
20. Give two examples of functional proteins.
21. What are amino acids?
22. What are standard amino acids?
23. What are non-standard amino acids?
24. What is a peptide bond?
25. Give two examples of polar amino acids.
26. Give two examples of non-polar amino acids.
27. What are the four structural levels of proteins?
28. What are two basic components of haemoglobin?
29. What is flagellin/tubulin/lamin/actin?
30. What is a nucleotide?
31. What is a nucleoside?
32. Which sugar is present in DNA/RNA?
33. Which four bases make the DNA/RNA?
34. What are the types of RNA?
35. What is ICTV?
36. When was Bergeys Manual first published?
37. How many volumes of Bergey's Manual of Systematic Bacteriology 'are published?

(B) Questions for Two Marks:
1. Define covalent / non-covalent / ester / Phospho-diester / Peptide / Glycosidic bond.
2. Define monosaccharide / disaccharide / oligosaccharide / polysaccharide with examples.
3. What are the classes of amino acids?
4. What are simple lipids / compound lipids / derived lipids. Give two examples.
5. What are phospholipids / glycolipids?
6. Describe primary /secondary / tertiary / quaternary structure of proteins.
7. Diagrammatically show the structure of DNA.
8. Draw the structure of Purine / Pyrimidine.
9. Explain base pairing in DNA / RNA.

10. Draw a purine base / pyrimidine base.
11. What are the advantages of classifying organisms?
12. What is species/genera/sub families/families/orders with respect to ICTV classification?

(C) Questions for Four Marks:

1. Describe the structure of Glucose / galactose / fructose / lactose.
2. Describe the structure of Starch / glycogen / peptidoglycan / chitin.
3. What are lipids? Give their biological importance.
4. Describe the structure of steroids/cholesterol.
5. What are amino acids? Give their classification with examples.
6. Describe simple lipids / compound lipids / derived lipids with suitable examples.
7. Describe triglycerides with examples.
8. Describe the four levels of protein structure.
9. With the help of diagram explain classification of lipids.
10. Describe structure and functions of flagellin / tubulin / lamin / actin.
11. Describe structure of Haemoglobin with the help of a diagram.
12. Explain the structure of DNA/RNA.
13. Explain the various types of RNA.
14. Describe structure and functions of rRNA.
15. Differentiate between DNA and RNA.
16. Give the different levels of viral classification with taxon suffixes for each level.
17. Explain the current grouping for the Bergey's Manual of Systematic Bacteriology.

(D) Fill in the Blanks:

1. Sucrose on hydrolysis gives one molecule of ------------- and ------------ each.
2. Ribose is a carbohydrate with the formula --------------.